ST. MARY'S COLLEGE OF MARYLAND
ST. MARY'S CITY, MARYLAND
41224

BANACH SPACES
of
ANALYTIC FUNCTIONS

PRENTICE-HALL SERIES IN MODERN ANALYSIS

R. CREIGHTON BUCK, editor

BANACH SPACES

of

ANALYTIC FUNCTIONS

KENNETH HOFFMAN

Department of Mathematics
Massachusetts Institute of Technology

PRENTICE-HALL, INC.,
Englewood Cliffs, N. J.

©—1962 by Prentice-Hall, Inc., Englewood Cliffs, N. J.

All rights reserved. No part of this book may be reproduced in any form, by mimeograph or any other means, without permission in writing from the publisher.

Library of Congress Catalog Card Number: 62-12452

Current printing (last digit):
12 11 10 9 8 7 6 5 4 3

PRINTED IN THE UNITED STATES OF AMERICA

05540—C

to Pat

PREFACE

There are not enough books which deal with the interplay between functional analysis and the theory of analytic functions. One reason for this is the fact that many of the techniques of functional analysis have a "real variable" character and are not directly applicable to problems which belong intrinsically to analytic function theory, e.g., problems of conformal mapping and Riemann surfaces. But there are parts of this theory which blend beautifully with the concepts and methods of functional analysis. These are fascinating areas of study for the general analyst, for three principal reasons: (a) the point of view of the algebraic analyst leads to the formulation of many interesting problems concerned with analytic functions; (b) when such problems are solved by a combination of the tools from the two disciplines, the depth of each discipline is increased; (c) the techniques of functional analysis often lend clarity and elegance to the proofs of classical theorems, and thereby make the results available in more general situations.

The main purpose of this monograph is to provide an introduction to the segment of mathematics in which functional analysis and analytic function theory merge successfully. Its spirit is close to that of abstract harmonic analysis, and, in fact, there is some overlap with the subject matter of harmonic analysis. Because this work is introductory, there has been no attempt to emulate either the depth of Zygmund's book on trigonometric series or the generality of the several books which treat harmonic analysis on groups. The subject matter is restricted to Banach spaces of analytic functions in the unit disc, roughly, those which are closely related to the Hardy spaces H^p ($1 \leq p \leq \infty$). The historical accounting sometimes falls a bit short of the mark. Some effort toward such an accounting is made in the sections entitled NOTES, at the end of each chapter. But a few relevant references have been omitted (for example, A. Taylor's papers in *Studia Mathematica*, 1950–51). The material is not discussed in its ultimate generality. Where proofs do carry over to more general contexts and the extension is not treated elsewhere, my method is usually to give the proofs in the unit disc and to discuss the generalizations afterward.

The first four chapters are devoted to the proofs of classical theorems on boundary-values and boundary integral representations for analytic functions in the unit disc which lie in the Hardy class H^p ($1 \leq p \leq \infty$). Some basic results on $(C, 1)$ summability of Fourier series are treated first, not because this is necessary, but because the reader who is not acquainted with approximate identity arguments may then see them in the context of Cesaro summability as well as in the context of Abel-Poisson summability. The treatment of Cesaro means first also helps to underscore the "real variable" nature of the basic propositions on boundary-values of H^p functions, i.e., to underscore the fact that the proofs do not utilize analyticity as such, but depend upon the fundamental theory of convergence, integration, and measure, plus a few basic facts about Banach spaces. The recent work of Helson and Lowdenslager has provided such "real variable" proofs for some of the F. and M. Riesz theorems on the space H^1, which originally leaned heavily on analytic function theory.

The fifth chapter contains the factorization theory for H^p functions, which, for its full strength, depends most decidedly on the fact that one is dealing with analytic functions of one complex variable. The chapter also contains a discussion of some partial extensions of the factorization, as well as a brief description of the classical approach to the theorems of the first five chapters.

There is a treatment of H^p spaces in a half-plane, which (for organizational reasons) occurs in Chapter 8. The principal facts are derived by reducing them to their counterparts in the unit disc. This is a bit unnatural, and it is done for two reasons: (i) to avoid a lengthy discussion of Fourier transforms, the natural tools for the study of the half-plane; (ii) to make available a detailed description of the relationship between H^p of the disc and H^p of the half-plane.

The remainder of the monograph deals with the structure of various Banach spaces and Banach algebras of analytic functions in the unit disc: H^p as a Banach space; the ideal theory of the algebra of continuous functions on the closed disc which are analytic in the interior; the invariant subspaces for the shift operator on the space H^2; the maximal ideal space of the algebra of bounded analytic functions in the disc. The material in this part of the book differs from that in the earlier part of the book, chiefly because the questions come from algebraic analysis. There is also an age difference; the bulk of the mathematics in the early part dates from 1925 or before, whereas most of the mathematics in the later chapters dates from 1949 to the present. The influence of Beurling's work is to be found throughout the latter part of the book, not only because many of the results are his, but also because he played a large role in reviving the functional analyst's interest in classical analytic function theory.

The level of the book is about that of the second-year graduate student. Chapter 1 summarizes the prerequisites for the reader, and these will carry him through most of the book; however, in the later chapters, some additional tools of analysis are used with only a reference for the proof, e.g., the Plancherel theorem, the Krein-Milman theorem, the existence of the Šilov boundary for a function algebra, and Šilov's theorem on the existence of idempotents in a Banach algebra. The prerequisites do not mention analytic functions, since the knowledge required in that area is elementary. The book contains one hundred exercises, with the usual dual purpose of exercises.

Thanks are due to many people for pointing out errors in the M.I.T. notes from which the monograph evolved, particularly to R. Askey, S. Bochner, H. Helson, G. Leibowitz, W. Rudin, and N. Starr. I want to thank the following people for the use of their unpublished results and/or manuscripts in the preparation of the book: R. Arens, H. Bear, E. Bishop, L. Carleson, A. Gleason, P. Halmos, H. Helson, D. Lowdenslager, D. Newman, W. Rudin, H. Shapiro, A. Shields, and J. Wermer. I especially want to thank R. Arens, I. Singer, and J. Wermer for their many helpful discussions.

Finally, for all their hard work, my gratitude goes to Mrs. Judith Bowers, who typed the bulk of the manuscript, and to the staff of Prentice-Hall, Inc.

Pacific Palisades, California   KENNETH HOFFMAN

CONTENTS

1. Preliminaries   1

    Measure and Integration 3
    Banach Spaces 5
    Hilbert Space and Fourier Series 8

2. Fourier Series   15

    Cesaro Means 16
    Characterization of Types of Fourier Series . . . 22
    Notes 24
    Exercises 25

3. Analytic and Harmonic Functions in the Unit Disc   27

    The Cauchy and Poisson Kernels 28
    Boundary Values 32
    Fatou's Theorem 34
    H^p Spaces 39
    Notes 39
    Exercises 40

4. The Space H^1 42

    The Helson-Lowdenslager Approach 42
    Szegö's Theorem 48
    Completion of the Discussion of H^1 50
    Dirichlet Algebras 54
    Notes 57
    Exercises 58

5. Factorization for H^p Functions   61

    Inner and Outer Functions 61
    Blaschke Products and Singular Functions 63
    The Factorization Theorem 67
    Absolute Convergence of Taylor Series 70
    Remarks on the Classical Approach 72

Functions of Bounded Characteristic 73
Notes 74
Exercises. 74

6. Analytic Functions with Continuous Boundary Values   77

Conjugate Harmonic Functions 78
Theorems of Fatou and Rudin 80
The Closed Ideals of A 82
Commutative Banach Algebras 89
Wermer's Maximality Theorem 93
Notes 95
Exercises. 95

7. The Shift Operator   98

The Shift Operator on H^2 98
More about Dirichlet Algebras 101
Invariant Subspaces for H^2 of the Half-plane 103
Isometries 108
The Shift Operator on L^2 111
The Vector-valued Case 114
Representations of H^∞ 116
Notes 119
Exercises. 119

8. H^p Spaces in a Half-plane   121

H^p of the Half-plane 121
Boundary Values for H^p Functions 124
The Paley-Wiener Theorem 131
Factorization for H^p Functions in a Half-plane 132
Notes 133
Exercises. 133

9. H^p as a Banach Space   136

Extreme Points 136
Isometries 142
Projections from L^p to H^p 149
Notes 155
Exercises. 156

10. H^∞ as a Banach Algebra   158

Maximal Ideals in H^∞ 159
Topological Structure of $\mathfrak{M}(H^\infty)$ 162
Discs in Fibers 166

L^∞ as a Banach Algebra 169
The Šilov Boundary 172
Inner Functions and the Šilov Boundary 175
Representing Measures and Annihilating Measures 180
Algebras on the Fibers 187
Maximality 193
Interpolation 194
Notes 206
Exercises 207

Bibliography 209

Index 215

CHAPTER 1

PRELIMINARIES

Measure and Integration

If X is a set, the collection of all subsets of X forms a ring, using the operations
$$A + B = (A \cup B) - (A \cap B)$$
$$AB = A \cap B.$$
A **σ-ring** of subsets of X is a subring of the ring of all subsets of X which is closed under the formation of countable unions (and, a fortiori, closed under the formation of countable intersections).

Suppose that X is a locally compact Hausdorff topological space, e.g., n-dimensional Euclidean space or a closed subset thereof. The **Baire subsets** of X are the members of the smallest σ-ring of subsets of X which contains every compact G_δ, i.e., every compact subset of X which is the intersection of a countable number of open sets. The **Borel subsets** of X are the members of the smallest σ-ring of subsets of X which contains every compact set. In Euclidean space, every compact (closed and bounded) set is a G_δ; hence, if X is a closed subset of Euclidean space, the Baire and Borel subsets of X coincide. When X is the real line or a closed interval on the line, the ring of Baire (Borel) subsets of X may also be described as the σ-ring generated by the half-open intervals $[a, b)$.

If X is a locally compact Hausdorff space, a **positive Baire (Borel) measure** on X is a function μ which assigns to every Baire (Borel) subset of X a non-negative real number (or $+\infty$), in such a way that
$$\mu\left(\bigcup_{n=1}^{\infty} A_n\right) = \sum_{n=1}^{\infty} \mu(A_n)$$
whenever $A_1, A_2, \ldots$ is a sequence of pairwise disjoint Baire (Borel) sets in X. The Borel measure μ is called **regular** if for each Borel set A
$$\mu(A) = \inf \mu(U)$$
the infimum being taken over the open sets U containing A. A Baire measure is always regular, and each Baire measure has a unique extension

1

to a regular Borel measure. For this reason (and others) we shall discuss only Baire measures on X.

The positive Baire measure μ is called **finite** if $\mu(A)$ is finite for each Baire set A. If X is compact, μ is finite if and only if $\mu(X)$ is finite.

Suppose X is the real line or a closed interval. Let F be a monotone increasing (non-decreasing) function on X which is continuous from the left:
$$F(x) = \sup_{t<x} F(t).$$
Define a function μ on semi-closed intervals $[a, b)$ by
$$\mu([a, b)) = F(b) - F(a).$$
Then μ has a unique extension to a positive Baire measure on X. The measure μ is finite if and only if F is bounded. If X is the real line, every positive Baire measure on X arises in this way from a left-continuous increasing function F. If X is a closed interval, a monotone function on X is necessarily bounded; thus, every finite positive Baire measure on X comes from such an increasing function. If X is either the line or an interval, the measure induced by $F(x) = x$ is called **Lebesgue measure**.

For the general locally compact X, a **Baire function** on X is a complex-valued function f on X such that $f^{-1}(S)$ is a Baire set for every Baire set S in the plane. Every continuous function is a Baire function. A **simple Baire function** for μ is a complex-valued function f on X of the form
$$f(x) = \sum_{k=1}^{n} \alpha_k \chi_{E_k}(x)$$
where

(i) $\alpha_1, \ldots, \alpha_n$ are complex numbers;
(ii) $E_1, \ldots, E_n$ are disjoint Baire sets of finite μ-measure;
(iii) χ_E denotes the characteristic function of the set E.

The simple functions form a vector space over the field of complex numbers. For such simple Baire functions f we define
$$\int f d\mu = \sum_{k=1}^{n} \alpha_k \mu(E_k).$$
If f is a simple function, so is $|f|$ and
$$\left| \int f d\mu \right| \leq \int |f| d\mu.$$

The Baire function f is called **integrable** with respect to μ if there exists a sequence of functions $\{f_n\}$ such that

(i) each f_n is a simple Baire function for μ;
(ii) $\lim_{m,n \to \infty} \int |f_m - f_n| d\mu = 0$;

(iii) f_n converges to f in measure; i.e., for each $\epsilon > 0$,
$$\lim_{n \to \infty} \mu(\{x; |f(x) - f_n(x)| \geq \epsilon\}) = 0.$$

If f is integrable, then for any such sequence $\{f_n\}$ the sequence $\{\int f_n d\mu\}$ converges and the limit of this sequence (which is independent of $\{f_n\}$) is denoted by $\int f d\mu$. Denote the class of μ-integrable functions by $L^1(d\mu)$. Then $L^1(d\mu)$ is a vector space and $f \to \int f d\mu$ is a linear functional on L^1. The Baire function f is in $L^1(d\mu)$ if and only if its real and imaginary parts are in $L^1(d\mu)$, or if and only if $|f|$ is in $L^1(d\mu)$. When f is in L^1,
$$\left|\int f d\mu\right| \leq \int |f| d\mu.$$

If f is a non-negative Baire function, one can always sensibly define $\int f d\mu$, so long as $+\infty$ is allowed as a value. That is, either f is integrable, or for every $K > 0$ there is a simple function $g \leq f$ with $\int g d\mu > K$. In the latter case, one defines $\int f d\mu = +\infty$.

A subset S of X has μ-**measure zero** if for each $\epsilon > 0$ there is a Baire set A containing S with $\mu(A) < \epsilon$. One can, if it is desirable, extend μ to the class of μ-**measurable sets**, such a set being one which differs from a Baire set by a set of measure zero. For our purposes, this will usually not be necessary. Any phenomenon which occurs except on a set of μ-measure zero is said to happen **almost everywhere** (relative to μ). One can also extend the concept of integrability to a function which agrees almost everywhere with a Baire function.

A basic theorem on integration is the **Lebesgue dominated convergence theorem.** If $\{f_n\}$ is a sequence of integrable functions such that the limit $f(x) = \lim_{n \to \infty} f_n(x)$ exists almost everywhere, and if there is a fixed integrable function g such that $|f_n| \leq |g|$ for each n, then f is integrable and
$$\int f d\mu = \lim_{n \to \infty} \int f_n d\mu.$$

Another basic fact is **Fubini's theorem,** a weak form of which is the following. Suppose μ is finite and f is a non-negative Baire function on the product space $X \times X$. If $f(x, y)$ is integrable in x for each fixed y and in y for each fixed x, then
$$\int [\int f(x, y) d\mu(x)] d\mu(y) = \int [\int f(x, y) d\mu(y)] d\mu(x).$$

If p is a positive number, the space $L^p(d\mu)$ consists of all Baire functions f such that $|f|^p$ is in $L^1(d\mu)$. If
$$f \in L^p(d\mu), \quad g \in L^q(d\mu), \quad \text{and} \quad \frac{1}{p} + \frac{1}{q} = 1$$
then $(fg) \in L^1(d\mu)$ and (**Hölder's inequality**)
$$\left|\int fg \, d\mu\right| \leq \left(\int |f|^p d\mu\right)^{1/p} \left(\int |g|^q d\mu\right)^{1/q}.$$

Let us note something about the spaces $L^p(d\mu)$ when X is compact and μ is a finite measure. In this case, every continuous function on X is integrable and the space of continuous functions is dense in L^1; i.e., if $f \in L^1$ and $\epsilon > 0$, there is a continuous g such that
$$\int |f - g| d\mu < \epsilon.$$
Also, if $p \geq 1$, then L^p is contained in L^1, and the continuous functions are a dense subspace of L^p:
$$\int |f - g|^p d\mu < \epsilon.$$

If μ_1 and μ_2 are positive Baire measures on X, we say that μ_1 is **absolutely continuous** with respect to μ_2 if every set of measure zero for μ_2 is a set of measure zero for μ_1. The **Radon-Nikodym theorem** states the following about finite measures: if μ_1 and μ_2 are finite, then μ_1 is absolutely continuous with respect to μ_2 if and only if
$$d\mu_1 = f d\mu_2$$
where f is some non-negative function in $L^1(d\mu_2)$. We say that μ_1 and μ_2 are **mutually singular** if there are disjoint Baire sets B_1 and B_2 such that
$$\mu_j(A) = \mu_j(A \cap B_j), \quad j = 1, 2,$$
for every Baire set A. The generalized **Lebesgue decomposition theorem** states the following: if μ_1 and μ_2 are any two finite positive Baire measures, then μ_1 is uniquely expressible in the form
$$\mu_1 = \mu_a + \mu_s$$
where μ_a is absolutely continuous with respect to μ_2, and μ_s and μ_2 are mutually singular. That is,
$$d\mu_1 = f d\mu_2 + d\mu_s$$
where $f \in L^1(d\mu_2)$, and μ_s and μ_2 are mutually singular. One usually calls f the derivative of μ_1 with respect to μ_2.

Let us look at this decomposition when X is a closed interval, and μ_2 is Lebesgue measure. Suppose μ is the positive measure determined by the increasing function F. Then, except on a set of Lebesgue measure zero, the function F is differentiable, and if $f = dF/dx$, then f is Lebesgue integrable and
$$d\mu = f dx + d\mu_s$$
where μ_s is mutually singular with Lebesgue measure. The latter means simply that μ_s is determined by an increasing function F_s such that $dF_s/dx = 0$ almost everywhere with respect to Lebesgue measure.

We wish to make a few brief comments about measures which assume arbitrary real or complex values. There are some technical difficulties here, but they do not arise if one treats only finite measures. Again, let X be a locally compact space. A **finite real Baire measure** on X is a countably additive and real-valued function μ on the class of Baire sets. One way to construct such a measure is to subtract two finite positive Baire measures: $\mu = \mu_1 - \mu_2$. The **Jordan decomposition theorem** states that this is the only example there is. Indeed, given such a real measure μ there are disjoint Baire sets B_1 and B_2 and finite positive measures μ_1 and μ_2 on B_1 and B_2, respectively, such that $\mu = \mu_1 - \mu_2$. This splitting (with B_1 and B_2 disjoint) is unique. The positive measure $\mu_1 + \mu_2$ is called the **total variation** of μ, denoted $|\mu|$. One defines absolute continuity and singularity of real measures using their total variations. It is then very easy to extend the decomposition into absolutely continuous and singular parts, for example, to the case where μ_1 is a real measure. If X is a closed interval on the real line, the finite real Baire measures on X are those induced by real-valued functions of bounded variation which are continuous from the left. The Jordan decomposition for such a measure corresponds to the canonical expression for a function of bounded variation as the difference of two increasing functions.

Finite complex Baire measures are defined similarly. If one wishes, such a measure μ is a function of the form $\mu_1 + i\mu_2$, where μ_1 and μ_2 are finite real Baire measures. Again, there are certain obvious extensions of some of the theorems above. And, of course, such a measure on a finite interval will be induced by a complex-valued function of bounded variation.

Banach Spaces

Let X be a real or complex vector space. A **norm** on X is a non-negative real-valued function $||\cdot \cdot \cdot||$ on X such that

(i) $||x|| \geq 0$; $||x|| = 0$ if and only if $x = 0$;
(ii) $||x + y|| \leq ||x|| + ||y||$;
(iii) $||\lambda x|| = |\lambda|\, ||x||$.

A real (complex) **normed linear space** is a real (complex) vector space X together with a specified norm on X. On such a space one has a metric ρ defined by

$$\rho(x, y) = ||x - y||.$$

If X is complete in this metric, we call X a **Banach space**. Completeness, then, means that if $\{x_n\}$ is a sequence of elements of X such that

$$\lim_{m,n \to \infty} ||x_m - x_n|| = 0$$

there exists an element x in X such that

$$\lim_{n \to \infty} ||x - x_n|| = 0.$$

Example 1. Let X be n-dimensional Euclidean space and define the norm of the n-tuple $x = (x_1, \ldots, x_n)$ by
$$||x||^2 = |x_1|^2 + \cdots + |x_n|^2.$$
Then X is a Banach space.

Example 2. Let S be a locally compact Hausdorff space and fix a positive Baire measure μ on S. Choose a number $p \geq 1$ and let $X = L^p(d\mu)$. Define the norm of $f \in L^p$ to be its L^p-norm
$$||f||_p = \left(\int |f|^p d\mu\right)^{1/p}.$$
On L^p as we have defined it, this is not a norm, since we may have $||f||_p = 0$ without $f = 0$. Consequently, we agree to identify henceforth two functions in $L^p(d\mu)$ which agree almost everywhere with respect to μ. Strictly speaking, then, the elements of $L^p(d\mu)$ will be equivalence classes of functions; however, we carry on with the same notation, simply identifying functions equal almost everywhere. With this convention the space $L^p(d\mu)$ ($p \geq 1$) is a Banach space using the L^p-norm. The crucial property of completeness says that if $\{f_n\}$ is a sequence of functions in L^p such that
$$\lim_{m,n\to\infty} \int |f_m - f_n|^p d\mu = 0$$
then there is an f in L^p such that $||f - f_n||_p \to 0$. The functions f_n do not necessarily converge pointwise to f; however, there is always a subsequence which converges to f almost everywhere. In this discussion we want to include the case $p = \infty$.

The space $L^\infty(d\mu)$ is simply the space of bounded Baire functions with the μ-essential sup norm:
$$||f||_\infty = \operatorname*{ess\,sup}_\mu{}_x |f(x)|$$
which means the infimum of $\sup_x |g(x)|$ as g ranges over all bounded Baire functions which agree with f almost everywhere with respect to μ. Of course, in all this discussion of $L^\infty(d\mu)$ we are identifying functions equal almost everywhere.

Example 3. Let S be a compact Hausdorff space and $X = C(S)$, the space of all continuous real (or complex) functions on S. Equip $C(S)$ with the sup (or uniform) norm
$$||f||_\infty = \sup_{x \in S} |f(x)|.$$
Then $C(S)$ is a Banach space.

Let X be a Banach space. We consider the space X^* of all linear functionals F on X which are continuous:
$$||x_n - x|| \to 0 \quad \text{implies} \quad |F(x_n) - F(x)| \to 0.$$

The set X^* forms a vector space in an obvious way. There is also a natural norm on X^*. It is based upon the observation that the linear functional F is continuous if and only if it is bounded; i.e., if and only if there is a constant $K > 0$ such that
$$|F(x)| \leq K||x||$$
for every x in X. The smallest such K is called the norm of F, i.e.,
$$||F|| = \sup_{||x|| \leq 1} |F(x)|.$$
With this norm X^* becomes a Banach space, the **conjugate space** of X.

Example 1. If X is Euclidean space, then every linear functional on X is continuous. Such a functional F has the form
$$F(x_1, \ldots, x_n) = a_1 x_1 + \cdots + a_n x_n$$
and
$$||F||^2 = |a_1|^2 + \cdots + |a_n|^2.$$

Example 2. Let S be a locally compact space and μ a positive Baire measure on S. Suppose $1 \leq p < \infty$ and that $X = L^p(d\mu)$. Then the conjugate space of X is $L^q(d\mu)$ where $\dfrac{1}{p} + \dfrac{1}{q} = 1$. If $p = 1$, $X^* = L^\infty(d\mu)$. If $g \in L^q(d\mu)$, then g induces a continuous linear functional F on L^p by
$$F(f) = \int f g \, d\mu, \qquad f \in L^p.$$
Every continuous linear functional on L^p has this form, and
$$||F|| = ||g||_q.$$
The conjugate space of $L^\infty(d\mu)$ contains $L^1(d\mu)$; but, except in trivial cases, it is larger than L^1.

Example 3. Let S be a compact Hausdorff space and $X = C(S)$, the space of continuous real (complex) functions on S. The conjugate space of $C(S)$ is the space of finite real (complex) Baire measures on S. This is the statement of the Riesz representation theorem. It arises as follows. Suppose μ is such a measure on S. The linear functional corresponding to μ is
$$F(f) = \int f \, d\mu, \qquad f \in C(S).$$
The norm of this functional F is called the **total variation** of μ on S. If μ is a real measure, the total variation of μ on S is simply $|\mu|(S)$, where $|\mu|$ denotes the measure known as the total variation of μ. If μ is complex, the total variation of μ on S is best thought of as the norm of the corresponding functional on $C(S)$, since the relation of this number to the total variations of the real and imaginary parts of μ is rather involved. Of course, if μ is a positive measure, the norm of F is simply $\mu(S)$. Needless to say, the important part of the Riesz theorem is the fact that given a bounded linear functional F on $C(S)$ there exists a finite measure μ such

that $F(f) = \int f d\mu$. This is proved by using the boundedness of F to extend F to the class of bounded Baire functions and then defining $\mu(E) = F(\chi_E)$ for each Baire set E.

Suppose X is a Banach space. One important property of continuous linear functionals on X is the **Hahn-Banach extension theorem**. If F is a bounded linear functional on a subspace Y of X, then F can be extended to a linear functional on X which has precisely the same bound (norm) as F.

In addition to the metric topology on the conjugate space X^*, we shall have occasion to consider another topology called the **weak-star topology** on X^*. It is defined as follows. Let $F_0 \in X^*$, and select a finite number of elements
$$x_1, \ldots, x_n \in X \quad \text{and} \quad \epsilon > 0.$$
Let
$$U = \{F \in X^*; |F(x_k) - F_0(x_k)| < \epsilon, k = 1, \ldots, n\}.$$
Such a set U is a basic weak-star neighborhood of F_0. A weak-star open set is any union of such basic neighborhoods U. We then have a topology on X^*. It is the weakest topology on X^* such that for each $x \in X$ the function $F \to F(x)$ is continuous on X^*. A topology on a set is, roughly, a scheme for deciding when two points are close together. In the weak-star topology two linear functionals are close together if their values on a finite number of elements of X are close together. In particular, a sequence $\{F_n\}$ converges to F in the weak-star topology if and only if
$$\lim_{n \to \infty} F_n(x) = F(x)$$
for each x in X.

We want the following basic result on X^* with the weak-star topology. If B is the closed unit ball in X^*:
$$B = \{F \in X^*; ||F|| \leq 1\}$$
then B is compact in the weak-star topology. This is a rather simple consequence of the fact that the Cartesian product of compact spaces is compact. We shall use this in the following way. If $\{F_n\}$ is a sequence of linear functionals on X with $||F_n|| \leq 1$, then this sequence has a weak-star cluster point in the unit ball; that is, there exists an $F \in X^*$ with $||F|| \leq 1$ such that $F(x)$ is a cluster point of the sequence $\{F_n(x)\}$ for every $x \in X$. For example, if $\{\mu_n\}$ is a sequence of positive Baire measures on the compact space S and if $\mu_n(S) \leq 1$ for each n, then there exists a finite measure μ such that $\int f d\mu$ is a cluster point of $\{\int f d\mu_n\}$ for every $f \in C(S)$.

Hilbert Space and Fourier Series

Let H be a real or complex vector space. An **inner product** on H is a function (,) which assigns to each ordered pair of vectors in H a scalar, in such a way that

(i) $(x_1 + x_2, y) = (x_1, y) + (x_2, y)$;
(ii) $(\lambda x, y) = \lambda(x, y)$;
(iii) $(y, x) = \overline{(x, y)}$;
(iv) $(x, x) \geq 0$; $(x, x) = 0$ if and only if $x = 0$.

Such a space H, together with a specified inner product on H, is called an **inner product space**. In any inner product space one has the **Cauchy-Schwarz inequality**:

$$|(x, y)|^2 \leq (x, x)(y, y).$$

This inequality is evident if $y = 0$. If $y \neq 0$, the inequality results from $0 \leq (x + \lambda y, x + \lambda y)$, where λ is the scalar

$$\lambda = -\frac{(x, y)}{(y, y)}.$$

From the Schwarz inequality it follows easily that $||x|| = (x, x)^{1/2}$ is a norm on H. If H is complete in this norm, we say that H is a **Hilbert space**. Thus, a Hilbert space is a Banach space in which the norm is induced by an inner product. By expanding $(x - y, x - y)$ and $(x + y, x + y)$ it is easy to see that the norm induced by an inner product satisfies the **parallelogram law**:

$$||x + y||^2 + ||x - y||^2 = 2(||x||^2 + ||y||^2).$$

Conversely, any such norm comes from an inner product. So, if one wishes, a Hilbert space is a Banach space in which the norm satisfies the parallelogram law.

Example 1. Let H be n-dimensional Euclidean space, and define the inner product of

$$x = (x_1, \ldots, x_n) \quad \text{and} \quad y = (y_1, \ldots, y_n)$$

by

$$(x, y) = x_1 \bar{y}_1 + \cdots + x_n \bar{y}_n.$$

Then H is a Hilbert space.

Example 2. Let X be a locally compact space and μ a positive Baire measure on X. Let $H = L^2(d\mu)$ with the inner product

$$(f, g) = \int f \bar{g} d\mu.$$

Then H is a Hilbert space.

The second example is the one we are interested in. For this space we already know one of the basic results about a Hilbert space H: every continuous linear functional on H is "inner product with some fixed vector in H"; that is, if F is a bounded linear functional on H, there is a unique vector y in H such that $F(x) = (x, y)$ for all x in H. The norm of F is $||F|| = ||y||$.

Two vectors x and y in H are called **orthogonal** if $(x, y) = 0$. If x and y are orthogonal, then
$$||x + y||^2 = ||x||^2 + ||y||^2.$$

Theorem. *Let S be a closed convex set in the Hilbert space H. Then S contains a unique element of smallest norm.*

Proof. Convexity means that if x and y are in S, so is $\lambda x + (1 - \lambda)y$ for any λ satisfying $0 \leq \lambda \leq 1$. Let $K = \inf_{x \in S} ||x||$. Choose a sequence $\{x_n\}$ of elements of S such that $\lim ||x_n|| = K$. Since S is convex, $\frac{1}{2}(x_m + x_n)$ is in S; so $||x_m + x_n|| \geq 2K$. Now the parallelogram law says:
$$||x_m + x_n||^2 + ||x_m - x_n||^2 = 2(||x_m||^2 + ||x_n||^2).$$
Since
$$\lim_{m,n} (||x_m||^2 + ||x_n||^2) = 2K^2 \quad \text{and} \quad ||x_m + x_n||^2 \geq 4K^2$$
we see that
$$\lim_{m,n} ||x_m - x_n|| = 0.$$
Since S is closed, the sequence $\{x_n\}$ converges to an element x in S. Obviously
$$||x|| = \lim ||x_n|| = K.$$
Furthermore, x is the only element in S of norm K. If y were another such element the sequence $x, y, x, y, \ldots$ would have to converge by the above argument.

If S is any collection of vectors in H, the **orthogonal complement** of S is the set $S^\perp$ of all vectors in H which are orthogonal to every vector in S. It is easy to see that $S^\perp$ is a closed subspace of H.

Theorem. *Let S be a closed subspace of H. Then $H = S \oplus S^\perp$; that is, every vector x in H is uniquely expressible in the form $x = y + z$ where y is in S and z is in $S^\perp$.*

Proof. Fix x_0 in H. Then, since S is a closed subspace,
$$x_0 - S = \{x_0 - y; y \text{ in } S\}$$
is easily seen to be a closed convex set in H. Let z_0 be the unique element of smallest norm in $x_0 - S$, say $z_0 = x_0 - y_0$ with y_0 in S. Claim z_0 is in $S^\perp$. Let y be in S. For any λ the vector $z_0 - \lambda y$ is in $x_0 - S$, so
$$||z_0 - \lambda y||^2 \geq ||z_0||^2.$$
If one takes
$$\lambda = \frac{(z_0, y)}{(y, y)},$$
one obtains $|(y, z_0)| \leq 0$; so $(y, z_0) = 0$. The uniqueness of y_0 and z_0 is a simple consequence of the disjointness of S and $S^\perp$.

The element y (above) is called the **orthogonal projection** of x into the closed subspace S. We see that y is simply the element in S closest to x.

Let N be any collection of vectors in H. We call N an **orthogonal set** if any two distinct vectors in N are orthogonal. An **orthonormal set** is an orthogonal set, each vector of which has norm 1.

Theorem. *Let* $N = \{x_1, \ldots, x_n\}$ *be a finite orthonormal set. For any vector* x *in* H, *the orthogonal projection of* x *into the subspace spanned by* N *is*

$$y = \sum_{k=1}^{n} (x, x_k) x_k.$$

Proof. Define y as above and put $z = x - y$. Then y is in the subspace spanned by $x_1, \ldots, x_n$ and, using the fact that $(x_i, x_j) = \delta_{ij}$, one sees that z is orthogonal to each x_k, hence is orthogonal to any linear combination of $x_1, \ldots, x_n$.

Corollary (Bessel's inequality). *If* $\{x_1, \ldots, x_n\}$ *is a finite orthonormal set, then for any vector* x *in* H

$$\sum_{k=1}^{n} |(x, x_k)|^2 \leq ||x||^2.$$

Equality holds if and only if x *is in the subspace spanned by* $x_1, \ldots, x_n$; *that is, if and only if*

$$x = (x, x_1)x_1 + \cdots + (x, x_n)x_n.$$

Proof. Write $x = y + z$ as above. Since $(y, z) = 0$,

$$||x||^2 = ||y||^2 + ||z||^2.$$

Using $(x_i, x_j) = \delta_{ij}$, one has

$$||y||^2 = \sum_{k=1}^{n} |(x, x_k)|^2.$$

These results can be extended to arbitrary orthonormal sets. For convenience we state them only for countable orthonormal sets.

Theorem. *Let* $\{x_n\}$ *be a countable orthonormal set of vectors in* H. *Let* x *be any vector in* H. *Then*

$$\sum_{n=1}^{\infty} |(x, x_n)|^2 \leq ||x||^2 \quad \text{(Bessel's inequality)}.$$

The sequence $s_n = \sum_{k=1}^{n} (x, x_k)x_k$ *converges to the orthogonal projection of* x *into the closed subspace spanned by* $\{x_n\}$. *Thus, the following are equivalent.*

(i) x *is in the closed subspace spanned by* $\{x_k\}$.

(ii) $||x||^2 = \sum_{n=1}^{\infty} |(x, x_n)|^2.$

(iii) $\lim_{n \to \infty} s_n = x.$

Proof. Let S_n be the closed subspace spanned by $\{x_1, \ldots, x_n\}$ and let S be the closed subspace spanned by the sequence $\{x_n\}$. Applying the last theorem to S_n, we have

$$\sum_{k=1}^{n} |(x, x_k)|^2 \leq ||x||^2$$

for every n. Thus the infinite series $\Sigma |(x, x_k)|^2$ converges, and its sum does not exceed $||x||^2$. If

$$s_n = (x, x_1)x_1 + \cdots + (x, x_n)x_n$$

then, with $n > m$, we have

$$||s_m - s_n||^2 = \sum_{k=m+1}^{n} |(x, x_k)|^2$$

and so $||s_m - s_n|| \to 0$ as $m, n \to \infty$. Let $y = \lim s_n$. It is easy to see that

$$(y, x_k) = \lim_{n \to \infty} (s_n, x_k) = (x, x_k)$$

for each k. Thus, the vector $z = x - y$ is orthogonal to each x_k, hence to S. Since y is in S, we see that y is the orthogonal projection of x in S. Now x is in S if and only if $x = y$. It is easy to see that

$$||y||^2 = \sum_{n=1}^{\infty} |(x, x_n)|^2.$$

Thus (i), (ii), and (iii) are equivalent.

When x (in the above theorem) is in the closed subspace spanned by $\{x_n\}$ one usually writes

$$x = \sum_{n=1}^{\infty} (x, x_n) x_n$$

for $\lim s_n = x$. Undoubtedly, the most important case of this last theorem is the one in which the closed subspace spanned by $\{x_n\}$ is all of H. The result then assumes this form:

Theorem. *Let* $N = \{x_n\}$ *be a countable orthonormal set in* H. *The following are equivalent.*

(i) N *is complete; that is, the only vector orthogonal to every* x_n *is the zero vector.*

(ii) N *is closed; that is, the closed subspace spanned by* N *is all of* H.

(iii) *For every* x *in* H,

$$\sum_{n=1}^{\infty} |(x, x_n)|^2 = ||x||^2.$$

(iv) *For every* x *in* H,

$$x = \sum_{n=1}^{\infty} (x, x_n) x_n.$$

Proof. Let S be the closed subspace spanned by $\{x_n\}$. Since
$$H = S \oplus S^{\perp},$$
we know that $S = H$ if and only if $S^{\perp} = \{0\}$. Thus (i) and (ii) are equivalent. The equivalence of (ii), (iii), and (iv) is contained in the last theorem.

Now let's take a look at the case we are interested in. Let $H = L^2(-\pi, \pi)$, the space of Lebesgue square-integrable functions on the closed interval $[-\pi, \pi]$ (complex values). The inner product is
$$(f, g) = \frac{1}{2\pi} \int_{-\pi}^{\pi} f(x)\overline{g(x)}dx.$$
In other words, $L^2(-\pi, \pi) = L^2(d\mu)$, where μ is the normalized Lebesgue measure $d\mu = \frac{1}{2\pi} dx$. Let $\varphi_n(x) = e^{inx}$. Then it is easy to verify that the set $\{\varphi_n\}_{n=-\infty}^{\infty}$ is an orthonormal set. This orthonormal set is complete. We assume this now and will prove it later. If $f \in L^2(-\pi, \pi)$, the numbers
$$c_n = (f, \varphi_n) = \frac{1}{2\pi} \int_{-\pi}^{\pi} f(x)e^{-inx}dx$$
are the **Fourier coefficients** of f. The formal series
$$\sum_{n=-\infty}^{\infty} c_n e^{inx}$$
is the **Fourier series** for f.

Our Hilbert space discussion above tells us the following. Suppose we start with $f \in L^2(-\pi, \pi)$ and define the Fourier coefficients c_n as above. Suppose $n \geq 0$ and we wish to approximate f in L^2-norm by a trigonometric polynomial
$$P(x) = \sum_{k=-n}^{n} a_k e^{ikx}.$$
Then the best such approximation is given by
$$s_n(x) = \sum_{k=-n}^{n} c_k e^{ikx}$$
that is, by taking $a_k = c_k$. We also know that the sequence of Fourier coefficients is square-summable and
$$\sum_{n=-\infty}^{\infty} |c_n|^2 = ||f||^2 = \frac{1}{2\pi} \int_{-\pi}^{\pi} |f(x)|^2 dx.$$
(Here we have used the completeness of $\{\varphi_n\}$). Furthermore,
$$f = \sum_{n=-\infty}^{\infty} c_n \varphi_n$$

i.e., f is the sum (in the Hilbert space sense) of its Fourier series. What this means, of course, is that the nth partial sum s_n of the Fourier series converges to f in the L^2-norm:

$$\lim_{n\to\infty} \frac{1}{2\pi} \int_{-\pi}^{\pi} |f(x) - s_n(x)|^2 dx = 0.$$

Note that we also have the **Riesz-Fischer theorem**: every square-summable sequence of complex numbers is the sequence of Fourier coefficients of a function in $L^2(-\pi, \pi)$. For if

$$\sum_{n=-\infty}^{\infty} |c_n|^2 < \infty$$

just put

$$s_n(x) = \sum_{k=-n}^{n} c_k e^{ikx}$$

and observe that $\{s_n\}$ converges in L^2 to a function f with Fourier coefficients c_n.

NOTES

For the preliminaries on measure and integration, some references are Halmos [38], Saks [79], Loomis [54], Titchmarsh [87], Dunford-Schwartz [25], Riesz-Nagy [73]. For the material on Banach spaces, see Banach [7], Loomis [54], Dunford-Schwartz [25], Riesz-Nagy [73]. The most convenient reference on Banach spaces is probably Loomis' book, since it has the essentials elegantly done. For the material on Hilbert spaces and orthonormal systems, see Zygmund [98], Stone [84], Riesz-Nagy [73], Halmos [37], Titchmarsh [87].

CHAPTER 2

FOURIER SERIES

Throughout this chapter we shall be working on the closed interval $[-\pi, \pi]$ on the real line. If f is a complex-valued Lebesgue-integrable function on that interval, the **Fourier coefficients** of f are the complex numbers

$$c_n = \frac{1}{2\pi} \int_{-\pi}^{\pi} f(x) e^{-inx} dx, \quad n = 0, \pm 1, \pm 2, \ldots$$

and the **Fourier series** for f is the formal series

$$\sum_{n=-\infty}^{\infty} c_n e^{inx}.$$

There are two fundamental questions about f and its associated series.

(1) Is f determined by its Fourier series?
(2) If so, how can we recapture f, given the Fourier series?

In asking the first question, we are treating f as an element of $L^1(-\pi, \pi)$; that is, we are identifying functions which differ only on a set of Lebesgue measure zero. Question 1, then, asks whether two integrable functions with the same sequence of Fourier coefficients agree almost everywhere. This question has an affirmative answer, as we shall soon see. Question 2 is a much meatier one, in part because it is stated in such a vague way. The first effort toward resolving Question 2 probably should be to form the **partial sums**

$$s_n(x) = \sum_{k=-n}^{n} c_k e^{ikx}$$

and to ask whether these functions s_n converge. Here one can ask whether the s_n converge pointwise, converge pointwise almost everywhere, converge uniformly, or converge in some type of norm. If they do converge, do they converge to f?

When f is square-integrable we have already seen that the partial sums converge to f in the L^2-norm (assuming the completeness of $\{e^{inx}\}$). One might hope that for f in L^1 the s_n converge to f in L^1-norm; however, this is not necessarily the case. One might hope that if f is continuous and

$f(-\pi) = f(\pi)$ then the s_n converge uniformly to f, but this fails. Indeed, for a continuous f it may happen that $\{s_n\}$ does not even converge pointwise. Really, no situation is quite as pleasant as the L^2 case, but this is not a hopeless roadblock. One simply looks for other ways to recapture f from its Fourier series. We shall look at one such method now. Before going on we should mention that for "smooth" functions f the partial sums s_n do converge pointwise, e.g., if f is of bounded variation. If f is, say, twice continuously differentiable, it is trivial to verify that $\{s_n\}$ converges uniformly, because two integrations by parts show that $c_n = O(1/n^2)$.

Cesaro Means

The (first) **Cesaro means** of the Fourier series for f are the arithmetic means

$$\sigma_n = \frac{1}{n}(s_0 + \cdots + s_{n-1}), \quad n = 1, 2, \ldots.$$

As we shall see, if f is in $L^p(-\pi, \pi)$, $1 \leq p < \infty$, then the Cesaro means σ_n converge to f in the L^p-norm. And if f is continuous [and $f(-\pi) = f(\pi)$] then the σ_n converge uniformly to f.

Now

$$s_n(x) = \sum_{k=-n}^{n} c_k e^{ikx}$$

$$= \sum_{k=-n}^{n} e^{ikx} \cdot \frac{1}{2\pi} \int_{-\pi}^{\pi} f(t) e^{-ikt} dt$$

$$= \frac{1}{2\pi} \int_{-\pi}^{\pi} f(t) \sum_{k=-n}^{n} e^{ik(x-t)} dt.$$

Thus

$$\sigma_n(x) = \frac{1}{2\pi} \int_{-\pi}^{\pi} f(t) K_n(x - t) dt$$

where $K_n(x)$ is the nth Cesaro mean of the series

$$\sum_{k=-\infty}^{\infty} e^{ikx}.$$

Thus

$$(n+1)K_{n+1}(x) - nK_n(x) = \sum_{k=-n}^{n} e^{ikx}$$

$$= \sum_{k=0}^{n} e^{ikx} + \sum_{k=1}^{n} e^{-ikx}$$

$$= \frac{1 - e^{i(n+1)x}}{1 - e^{ix}} + \frac{1 - e^{-i(n+1)x}}{1 - e^{-ix}} - 1$$

$$= \frac{\cos nx - \cos(n+1)x}{1 - \cos x}.$$

Since $K_1(x) = 1$, it is easy to see that
$$K_n(x) = \frac{1}{n}\left[\frac{1-\cos nx}{1-\cos x}\right]$$
$$= \frac{1}{n}\left[\frac{\sin \frac{n}{2} x}{\sin \frac{1}{2} x}\right]^2.$$

This sequence of functions K_n is called **Fejer's kernel**. We have shown for any integrable f on $[-\pi, \pi]$ that the nth Cesaro mean of the Fourier series for f is
$$\sigma_n(x) = \frac{1}{2\pi}\int_{-\pi}^{\pi} f(t)K_n(x-t)dt$$
where K_n is Fejer's kernel. Here are some properties of K_n.

(i) $K_n \geq 0$

(ii) $\dfrac{1}{2\pi}\displaystyle\int_{-\pi}^{\pi} K_n(x)dx = 1$

(iii) If I is any open interval about $x = 0$, then
$$\limsup_{n\to\infty,\, x\notin I} |K_n(x)| = 0 \quad (|x| \leq \pi).$$

Property (i) is evident from the derived expression for K_n. Property (ii) simply states that the nth Cesaro mean of the Fourier series for the constant function 1 is 1. Property (iii) results from a few simple inequalities. If $0 < \delta < \pi$ and if $\pi \geq |x| \geq \delta$, then
$$(\sin \tfrac{1}{2} x)^2 \geq (\sin \tfrac{1}{2} \delta)^2$$
so
$$|K_n(x)| \leq \frac{1}{n(\sin \tfrac{1}{2}\delta)^2}, \quad \text{for } \delta \leq |x| \leq \pi.$$

Now K_n is also an even function, but we shall make no use of that fact. All that we want to know about Cesaro means will be proved using only the above three properties of Fejer's kernel.

Any sequence of Lebesgue-integrable functions K_n which possesses properties (i), (ii), and (iii) above we shall call an **approximate identity** (for L^1). (Some call this a positive kernel.) We shall comment on the terminology later. We shall also see other approximate identities later. As one example slightly different from Fejer's kernel K_n, one might take
$$K'_n = \begin{cases} 2K_n, & 0 \leq x \leq \pi \\ 0, & -\pi \leq x < 0. \end{cases}$$

Theorem. *Let f be a function in $L^p(-\pi, \pi)$, where $1 \leq p < \infty$. Then the Cesaro means of the Fourier series for f converge to f in the L^p-norm. If f is* continuous and $f(-\pi) = f(\pi)$, then the Cesaro means converge uniformly to f.

Proof. Now
$$\sigma_n(x) = \frac{1}{2\pi}\int_{-\pi}^{\pi} f(t)K_n(x-t)dt.$$
If we extend f to a function on the real line which is periodic with period 2π, this may be written
$$\sigma_n(x) = \frac{1}{2\pi}\int_{-\pi}^{\pi} f(x-t)K_n(t)dt.$$
The periodicity condition $f(-\pi) = f(\pi)$ is not important for f in L^p, since we only know $f(x)$ almost everywhere; however, in discussing a continuous f it is important because it makes the periodic extension of f continuous. Let us examine the continuous case first. Since $\int K_n = 1$,
$$\sigma_n(x) - f(x) = \frac{1}{2\pi}\int_{-\pi}^{\pi} [f(x-t) - f(x)]K_n(t)dt.$$
If $\delta > 0$, we write
$$\sigma_n(x) - f(x) = \frac{1}{2\pi}\int_{-\delta}^{\delta} [f(x-t) - f(x)]K_n(t)dt$$
$$+ \frac{1}{2\pi}\int_{|t|\geq \delta} [f(x-t) - f(x)]K_n(t)dt$$
and we see that
$$|\sigma_n(x) - f(x)| \leq \sup_{-\delta<t<\delta} |f(x-t) - f(x)| + 2||f||_\infty \cdot \sup_{|t|\geq \delta} K_n(t).$$
If f is continuous at x and δ is small, the number $|f(x-t) - f(x)|$ is small for $|t| \leq \delta$; and since
$$\lim_{n\to\infty} \sup_{|t|\geq \delta} K_n(t) = 0,$$
we see that
$$\lim_{n\to\infty} \sigma_n(x) = f(x).$$
If f is continuous on any closed interval $a \leq x \leq b$, then f is uniformly continuous there and it is easily seen that $\sigma_n(x) \to f(x)$ uniformly on $[a, b]$.

For f in L^p we wish to estimate
$$||\sigma_n - f||_p.$$
Let g be any function in L^q, where $\frac{1}{p} + \frac{1}{q} = 1$. Then
$$\frac{1}{2\pi}\int_{-\pi}^{\pi} [\sigma_n(x) - f(x)]g(x)dx = \frac{1}{4\pi^2}\iint [f(x-t) - f(x)]g(x)K_n(t)dxdt$$
and thus

$$\left|\frac{1}{2\pi}\int_{-\pi}^{\pi}[\sigma_n(x) - f(x)]g(x)dx\right|$$

$$\leq \frac{1}{2\pi}\int_{-\pi}^{\pi}\left|\frac{1}{2\pi}\int_{-\pi}^{\pi}[f(x-t) - f(x)]g(x)dx\right|K_n(t)dt.$$

Using the Hölder inequality, the inside integral is not larger in modulus than

$$||g||_q \cdot ||f_t - f||_p$$

where $f_t(x) = f(x - t)$. Thus

$$\left|\frac{1}{2\pi}\int_{-\pi}^{\pi}[\sigma_n(x) - f(x)]g(x)dx\right| \leq ||g||_q \cdot \frac{1}{2\pi}\int_{-\pi}^{\pi}||f_t - f||_p K_n(t)dt$$

for every $g \in L^q$. Therefore,

$$||\sigma_n - f||_p \leq \frac{1}{2\pi}\int_{-\pi}^{\pi}||f_t - f||_p K_n(t)dt.$$

For, since L^q is the conjugate space of L^p, if we are given a function h in L^p we can also find (by the Hahn-Banach theorem) a g in L^q such that $||g||_q = 1$ and $\int hg = ||h||_p$.

Now if $\delta > 0$, write

$$||\sigma_n - f||_p \leq \frac{1}{2\pi}\int_{-\delta}^{\delta}||f_t - f||_p K_n(t)dt + \frac{1}{2\pi}\int_{|t|\geq\delta}||f_t - f||_p K_n(t)dt$$

$$\leq \sup_{-\delta<t<\delta}||f_t - f||_p + 2||f||_p \cdot \sup_{|t|\geq\delta} K_n(t).$$

If δ is small, $||f_t - f||_p$ is small for $|t| \leq \delta$, i.e., translation is continuous in the L^p-norm. Thus

$$\lim_{n\to\infty}||\sigma_n - f||_p = 0.$$

Theorem. *If f is in $L^\infty(-\pi, \pi)$ then the Cesaro means of the Fourier series for f converge to f in the weak-star topology on L^∞.*

Proof. As we observed above, for any g in L^1

$$\left|\frac{1}{2\pi}\int_{-\pi}^{\pi}[\sigma_n(x) - f(x)]g(x)dx\right| \leq \sup_{-\delta<t<\delta}\left|\frac{1}{2\pi}\int_{-\pi}^{\pi}[f(x-t) - f(x)]g(x)dx\right|$$

$$+ 2||f||_\infty \cdot \sup_{|t|\geq\delta} K_n(t).$$

Thus, we need only prove that

$$\lim_{t\to 0}\int_{-\pi}^{\pi}[f(x-t) - f(x)]g(x)dx = 0$$

or that

$$\lim_{t\to 0}\int_{-\pi}^{\pi}f(y)[g(y) - g(y-t)]dt = 0.$$

This follows from the fact that f is bounded and $\|g - g_t\|_1 \to 0$. So

$$\frac{1}{2\pi}\int_{-\pi}^{\pi} \sigma_n(x)g(x)dx \to \frac{1}{2\pi}\int_{-\pi}^{\pi} f(x)g(x)dx$$

for every g in L^1; i.e., $\sigma_n \to f$ in the weak-star topology.

We should perhaps comment that the analogue of this last theorem holds for measures as well. If μ is any finite complex Baire measure on $[-\pi, \pi]$ we can define the Fourier coefficients

$$c_n = \int_{-\pi}^{\pi} e^{-inx} d\mu(x)$$

and the associated Fourier series. The coefficients c_n are often called the Fourier-Stieljes coefficients of the measure. We would not expect the Cesaro means for μ to converge as functions, but we might expect the measures $\frac{1}{2\pi}\sigma_n(x)dx$ to converge to μ in the weak-star topology on measures. This is the case, but the measure μ must have period 2π. All this means is that if μ has a "point mass" at π or $-\pi$ these masses must be the same: $\mu(\{-\pi\}) = \mu(\{\pi\})$. A better way to formulate this condition is that μ is really a measure on the circle obtained by identifying $-\pi$ and π.

Theorem. Let μ be a finite (periodic) complex Baire measure on the interval $[-\pi, \pi]$ and let σ_n be the nth Cesaro mean of the Fourier series for μ. If f is any continuous function of period 2π, then

$$\lim_{n \to \infty} \frac{1}{2\pi}\int_{-\pi}^{\pi} f(x)\sigma_n(x)dx = \int_{-\pi}^{\pi} f(x)d\mu(x)$$

that is, the measures $\frac{1}{2\pi}\sigma_n dx$ converge to μ in the weak-star topology.

Proof.

$$\frac{1}{2\pi}\int_{-\pi}^{\pi} f(x)\sigma_n(x)dx = \int_{-\pi}^{\pi}\left[\frac{1}{2\pi}\int_{-\pi}^{\pi} f(x)K_n(x-t)dx\right]d\mu(t)$$

$$= \int_{-\pi}^{\pi} \tau_n(t)d\mu(t)$$

where τ_n is the nth Cesaro mean for f. Since $\tau_n \to f$ uniformly, we are done.

One of the corollaries to the sequence of theorems above is **Fejer's theorem**: every continuous function of period 2π is a uniform limit of trigonometric polynomials

$$p(x) = \sum_{k=-n}^{n} a_k e^{ikx}.$$

From this it follows that the orthonormal family $\{e^{inx}\}$ is complete in $L^2(-\pi, \pi)$, for the closed linear span of these functions contains the continuous functions, which are dense in L^2. Of course, the completeness is also contained in the result that for f in L^2 the σ_n converge to f in L^2.

We also know now that every integrable function is determined by its sequence of Fourier coefficients; indeed, we know that any periodic measure is determined by its Fourier coefficients.

Some comments may be in order, to place the results about Cesaro means in proper perspective. Having defined Fourier coefficients for, say, Lebesgue integrable functions, it is clear that if we add two functions the respective Fourier coefficients add. For f, g in $L^1(-\pi, \pi)$ we can also define a multiplication (though not pointwise). The multiplication we have in mind is **convolution**:

$$(f*g)(x) = \frac{1}{2\pi} \int_{-\pi}^{\pi} f(x-t)g(t)dt.$$

Using the Fubini theorem, it is easy to see that $f*g$ is again in L^1 and that

$$||f*g||_1 \leq ||f||_1 \, ||g||_1.$$

Also, one can see that convolution is associative and makes L^1 into a linear algebra. The nth Fourier coefficient of $f*g$ is the product of the nth Fourier coefficients of f and g:

$$\frac{1}{2\pi}\int_{-\pi}^{\pi} e^{-inx}(f*g)(x)dx = \frac{1}{2\pi}\int_{-\pi}^{\pi} e^{-inx} \cdot \frac{1}{2\pi}\int_{-\pi}^{\pi} f(x-t)g(t)dtdx$$

$$= \frac{1}{2\pi}\int_{-\pi}^{\pi} g(t) \cdot \left[\frac{1}{2\pi}\int_{-\pi}^{\pi} e^{-inx}f(x-t)dx\right] dt$$

$$= \frac{1}{2\pi}\int_{-\pi}^{\pi} g(t)e^{-int}dt \cdot \frac{1}{2\pi}\int_{-\pi}^{\pi} e^{-iny}f(y)dy.$$

One can also define the convolution of two measures. Let us do this only in case one of the measures is absolutely continuous with respect to Lebesgue measure, i.e., has the form $\frac{1}{2\pi}f(x)dx$ with f in L^1. The convolution of f and μ is the function

$$(f*\mu)(x) = \int_{-\pi}^{\pi} f(x-t)d\mu(t).$$

(The Fubini theorem is required to see that this definition makes sense.) Again, it is easy to verify that the Fourier coefficients of $f*\mu$ are the products of the corresponding coefficients for f and μ. An important case is the one in which μ is the **Dirac delta measure**, i.e., the point mass at 0. This measure δ_0 assigns the measure 1 to a Baire set if it contains the point 0, and otherwise assigns the measure 0. If f is in L^1 then $f*\delta_0 = f$; i.e., δ_0 serves as an identity under convolution. This corresponds to the fact that the Fourier coefficients of δ_0 are all equal to 1.

If f is in L^1 the Cesaro means for f converge to f in L^1. This is because $\sigma_n = f*K_n$ and the measures $\frac{1}{2\pi}K_n(x)dx$ are approaching the delta measure

δ_0. This is why we call $\{K_n\}$ an approximate identity for L^1. Of course, the Fejer kernel K_n is just the nth Cesaro mean of the Fourier series for the delta measure δ_0.

The results above hold when $\{K_n\}$ is any approximate identity for L^1. That is, $\{f*K_n\}$ converges uniformly to f if f is continuous, converges to f in L^p-norm if $1 \leq p < \infty$, converges weak-star to f if f is in L^∞, and $\{\mu*K_n\}$ converges to μ weak-star if μ is a measure on the circle. The proofs are exactly the same as those above if each K_n is a bounded function. This will be the case in the approximate identities we consider. If the K_n are not bounded, one must first verify that the convolution of an L^1 function and an L^p function is in L^p, and then the proofs proceed as above.

Characterization of Types of Fourier Series

To complete our preliminary discussion of Fourier series, we turn to the following question. Suppose we are given a formal Fourier series

$$\sum_{n=-\infty}^{\infty} c_n e^{inx}.$$

How can we tell whether this is the Fourier series of an L^1 function? An L^p function? A measure? A continuous function? For L^2 we know the answer: the sequence $\{c_n\}$ must be square-summable. Certain rough tests can be applied in the other cases. For example, the sequence of Fourier coefficients of any finite measure must be bounded (by the total variation of the measure on $[-\pi, \pi]$). This includes the case of an absolutely continuous measure $\frac{1}{2\pi} f(x) dx$, f in L^1. For this case one can say even more: the Fourier coefficients of an integrable function tend to zero:

$$\lim_{|n| \to \infty} |c_n| = 0.$$

This is the Riemann-Lebesgue lemma, and it is not difficult to prove. For instance, one can prove it first when f is the characteristic function of an interval $[a, b]$. Then one obtains

$$c_n = \frac{1}{2\pi} \int_a^b e^{-inx} dx$$

$$= \frac{i}{2\pi n} [e^{-inb} - e^{-ina}]$$

so that

$$|c_n| \leq \frac{1}{\pi |n|}.$$

The result then follows for step functions, i.e., linear combinations of characteristic functions of intervals. Since the step functions are dense in

L^1 one has the general result. Of course the Fourier coefficients of a measure need not tend to zero.

A reasonably satisfactory answer to our question about a formal series can be given in terms of the Cesaro means of the series.

Theorem. *A formal Fourier series is the Fourier series of*
  (i) *an L^p function, $1 < p \leq \infty$;*
  (ii) *an L^1 function;*
  (iii) *a continuous function of period 2π;*
  (iv) *a finite measure;*
  (v) *a finite positive measure;*
if, and only if, the Cesaro means σ_n
  (i)' *are bounded in L^p-norm;*
  (ii)' *converge in the L^1-norm;*
  (iii)' *converge uniformly;*
  (iv)' *are bounded in L^1-norm;*
  (v)' *are each non-negative.*

Proof. We have already proved most of the implications $k \to k'$, and the rest are easy to fill in. For example, if σ_n is the nth Cesaro mean of the Fourier series of a finite real measure, then

$$\sigma_n(x) = \int K_n(x-t) d\mu(t)$$

and so

$$\frac{1}{2\pi} \int_{-\pi}^{\pi} |\sigma_n(x)| dx \leq \frac{1}{2\pi} \int_{-\pi}^{\pi} \int_{-\pi}^{\pi} K_n(x-t) d|\mu|(t) dx$$
$$= |\mu|([-\pi, \pi]).$$

For complex measures take real and imaginary parts.

So all we need prove is that if the σ_n satisfy a condition k' then we have a Fourier series of type k. First let us make an observation about any formal series:

$$\lim_{n \to \infty} \frac{1}{2\pi} \int_{-\pi}^{\pi} e^{-imx} \sigma_n(x) dx = c_m.$$

For if $n > |m|$ the mth Fourier coefficient of σ_n is $\dfrac{n - |m|}{n} c_m$.

Suppose the Cesaro means are bounded in L^p-norm, where $1 < p \leq \infty$. We may as well assume that

$$||\sigma_n||_p \leq 1, \quad n = 1, 2, 3, \ldots.$$

The σ_n then lie in the unit ball of the conjugate space of L^q, where $\dfrac{1}{p} + \dfrac{1}{q} = 1$. Since this unit ball is weak-star compact, there is a function f in L^p with $||f||_p \leq 1$ such that every weak-star neighborhood of f contains σ_n for infinitely many values of n. In other words, given any g in L^q the numbers $\int \sigma_n g dx$ are near $\int f g dx$ for infinitely many values of n. Each e^{imx}

is in L^q; since the mth Fourier coefficient of σ_n tends to c_m, it must be that c_m is the mth Fourier coefficient of f. This takes care of (i).

If the σ_n converge in L^1 norm, they converge in that norm to an integrable function f. Since

$$\left|\frac{1}{2\pi}\int_{-\pi}^{\pi}[f(x)-\sigma_n(x)]e^{-imx}dx\right| \leq ||f-\sigma_n||_1$$

c_m is the mth Fourier coefficient of f.

If the σ_n converge uniformly, they converge to a continuous f, and this f has the desired Fourier coefficients by a similar argument.

Suppose

$$||\sigma_n||_1 \leq 1, \quad n = 1, 2, 3, \ldots.$$

Then the measures $d\mu_n = \frac{1}{2\pi}\sigma_n(x)dx$ are bounded in total variation by 1. The space of measures is the conjugate space of the Banach space of continuous functions. The μ_n all lie in the unit ball of this conjugate space; hence, they have a weak-star cluster point μ. Since each e^{imx} is continuous, the same sort of argument used above shows that μ has the desired Fourier coefficients.

Suppose that

$$\sigma_n(x) \geq 0, \quad n = 1, 2, \ldots.$$

Then

$$||\sigma_n||_1 = \frac{1}{2\pi}\int_{-\pi}^{\pi}\sigma_n(x)dx = c_0.$$

Thus the σ_n are bounded in L^1-norm. By the last result, our series is the Fourier series of a finite (periodic) measure μ. So μ is the weak-star limit of the measures $\frac{1}{2\pi}\sigma_n(x)dx$; i.e.,

$$\lim_{n\to\infty}\frac{1}{2\pi}\int_{-\pi}^{\pi}g(x)\sigma_n(x)dx = \int_{-\pi}^{\pi}g(x)d\mu(x)$$

for every continuous g (of period 2π). If $g \geq 0$, so is $g\sigma_n$; so $\int g d\mu \geq 0$. Thus μ is a positive measure.

NOTES

The chief reference on Fourier series is Zygmund's book [98]. It should be consulted for the expansion and extension of the results on Fourier series. One may also consult the books by Titchmarsh [87] and Rudin [76]. Fejer's theorem is in [28]. The characterizations of Fourier series of various types are due to Steinhaus [83] and Gross [36] for L^1, G. C. and W. H. Young [97] for L^p, $p > 1$, Young [96] for measures, Herglotz [45] for positive measures. See also Carathéodory [16] for Fourier series of increasing functions (positive measures).

EXERCISES

1. The series $\sum_{n=1}^{\infty} \frac{1}{n} \sin nx$ is the Fourier series of an L^2 function. Which L^2 function?

2. Let f be a Baire function on $[-\pi, \pi]$ and suppose $|f(x)| \leq 1$. Prove that each Cesaro mean of the Fourier series for f satisfies $|\sigma_n(x)| \leq 1$. If for some n and x we have $|\sigma_n(x)| = 1$, then f is constant.

3. Let Σa_n be an infinite series. Suppose the Cesaro means of the series converge to some number a, and suppose also that $a_n = O\left(\frac{1}{n}\right)$. Prove that the partial sums also converge to a. (G. H. Hardy [40]).

4. For an integrable function on $[-\pi, \pi]$, write the partial sums of the Fourier series for f in the form

$$s_n(x) = \frac{1}{2\pi} \int_{-\pi}^{\pi} f(t) D_n(x-t) dt;$$

that is, determine the (Dirichlet) kernel D_n explicitly. Use the result of Exercise 3 to prove that, when f is of bounded variation,

$$\lim_{n \to \infty} s_n(x) = \tfrac{1}{2} \lim_{t \to 0^+} [f(x+t) + f(x-t)].$$

5. If

$$K_n(x) = \begin{cases} n, & |x| \leq \pi/n \\ 0, & \pi/n \leq |x| \leq \pi, \end{cases}$$

verify that $\{K_n\}$ is an approximate identity for L^1. What, specifically, do the various convergence theorems of this chapter say for this particular approximate identity?

6. If $p \geq 1$, prove that the convolution of an L^1 function and an L^p function is in L^p. If $p = \infty$, prove that the convolution is continuous.

7. Give an example of two distinct measures on $[-\pi, \pi]$ which have the same Fourier series if we do not identify $-\pi$ and π.

8. If f is in L^2, prove that

$$g(x) = \frac{1}{2\pi} \int_{-\pi}^{\pi} f(x+t) f(t) dt$$

is continuous. How does the Fourier series of g behave?

9. For finite measures μ_1 and μ_2 define the convolution $\mu_1 * \mu_2$ to be the unique measure which yields the linear functional L on the continuous functions:

$$L(f) = \iint f(x+y) d\mu_1(x) d\mu_2(y).$$

Now prove that, if μ_1 is absolutely continuous with respect to Lebesgue measure, then $\mu_1 * \mu_2$ is also, and that its derivative is given by the convolution formula used in this chapter to define the convolution of a function and a measure.

10. Use Fejer's theorem to prove the Weierstrass approximation theorem: On a closed interval of the real line, every continuous function is a uniform limit of polynomials.

11. Prove that the partial sums of the Fourier series of a function of bounded variation are uniformly bounded.

CHAPTER 3

ANALYTIC AND HARMONIC FUNCTIONS IN THE UNIT DISC

Let D denote the open unit disc in the complex plane:
$$D = \{z; |z| < 1\}$$
and let C denote the unit circle:
$$C = \{z; |z| = 1\}.$$
Recall that a complex-valued function f is **analytic** in D provided that it is the sum of a convergent power series
$$f(z) = \sum_{n=0}^{\infty} a_n z^n.$$
This just means that f has a derivative at each point of D. A complex-valued function u on D is **harmonic** if it satisfies Laplace's equation:
$$\frac{\partial^2 u}{\partial x^2} + \frac{\partial^2 u}{\partial y^2} = 0.$$
Any analytic function is a complex-valued harmonic function. A real-valued function u is harmonic if and only if it is the real part of an analytic function, $f = u + iv$. For a real-valued harmonic u, any v such that $u + iv$ is analytic is called a **harmonic conjugate** of u. Such a v is just a real-valued function which with u satisfies the Cauchy-Riemann equations:
$$\frac{\partial u}{\partial x} = \frac{\partial v}{\partial y}, \qquad \frac{\partial u}{\partial y} = -\frac{\partial v}{\partial x}.$$
The harmonic conjugate of u is unique up to an additive constant. In other words, given the real harmonic function u, there is a unique real harmonic function v which is conjugate to u and vanishes at the origin.

The Cauchy and Poisson Kernels

If f is an analytic (or harmonic) function in the unit disc D, we wish to inquire when f has boundary values, and how f is determined by its boundary values. Roughly, then, we shall investigate conditions under which the limits

$$f(\theta) = \lim_{r \to 1} f(re^{i\theta})$$

exist and define a function on the unit circle C. Then we shall ask how f is determined by this function on the circle. If f is actually analytic in a disc of radius $1 + \epsilon$, certainly f has boundary values and is determined by these boundary values in accordance with the Cauchy integral formula:

$$f(z) = \frac{1}{2\pi i} \int_C \frac{f(\xi)}{\xi - z} d\xi.$$

For our purposes, it will be more convenient to write the Cauchy formula in the form

$$f(z) = \frac{1}{2\pi} \int_{-\pi}^{\pi} f(e^{i\theta}) \cdot \frac{e^{i\theta}}{e^{i\theta} - z} d\theta.$$

If f is merely harmonic in a disc of radius $1 + \epsilon$, we do not have a Cauchy integral formula; however, we can recapture f from its boundary values by means of the Poisson integral formula. Both of these formulas for the disc are intimately related to Fourier series; before we relate the Poisson formula, let us establish the relationship between harmonic functions and Fourier series. We shall not give the briefest discussion possible. Instead, let us roam around a bit in order to acquire a feeling for what is going on.

First, suppose f is analytic in the open disc:

$$f(z) = \sum_{n=0}^{\infty} a_n z^n.$$

Let $f_r(\theta) = f(re^{i\theta})$. For a fixed r, f_r is a function defined on the unit circle; i.e., if we restrict f to the circle of radius r, we obtain a continuous function on that circle which we can also interpret as a function on the unit circle. Now

$$f_r(\theta) = \sum_{n=0}^{\infty} a_n r^n e^{in\theta}.$$

That is, the nth Fourier coefficient of f_r is $a_n r^n$, $n \geq 0$, and is zero for $n < 0$. If f is analytic in the closed disc, the boundary value function f_1 has the Fourier coefficients a_n. Let's look at the Cauchy formula from this point of view:

$$f_r(\theta) = \frac{1}{2\pi} \int_{-\pi}^{\pi} f(e^{it}) \frac{e^{it}}{e^{it} - re^{i\theta}} dt.$$

If we agree to write $f(t)$ for $f(e^{it})$, we have

$$f_r(\theta) = \frac{1}{2\pi} \int_{-\pi}^{\pi} f(t) \frac{1}{1 - re^{i(\theta-t)}} dt$$

$$= \frac{1}{2\pi} \int_{-\pi}^{\pi} f(t) C_r(\theta - t) dt$$

where

$$C_r(\theta) = \frac{1}{1 - re^{i\theta}}.$$

In other words, f_r is the convolution $f_r = f * C_r$, provided f here denotes f on the unit circle. Thus, the Fourier coefficients of f_r are the products of the Fourier coefficients of $f(e^{it})$ and those of C_r:

$$f(e^{i\theta}) = \sum_{n=0}^{\infty} a_n e^{in\theta}$$

$$C_r(e^{i\theta}) = \sum_{n=0}^{\infty} r^n e^{in\theta}$$

$$f_r(\theta) = \sum_{n=0}^{\infty} a_n r^n e^{in\theta}.$$

Suppose u is harmonic (and real-valued) in the disc. Then u is the real part of an analytic function, or

$$u(z) = f(z) + \overline{f(z)}$$

where f is analytic. If

$$f(z) = \sum_{n=0}^{\infty} a_n z^n$$

then

$$u(z) = 2 \operatorname{Re} a_0 + \sum_{n=1}^{\infty} a_n z^n + \sum_{n=1}^{\infty} \bar{a}_n \bar{z}^n.$$

If we restrict u to the circle of radius r,

$$u_r(\theta) = u(re^{i\theta}) = \sum_{n=-\infty}^{\infty} c_n r^{|n|} e^{in\theta}$$

where $c_0 = 2 \operatorname{Re} a_0$, $c_n = a_n$ for $n > 0$ and $c_n = \bar{a}_{-n}$ for $n < 0$. If u is harmonic in the closed disc, then the boundary function u_1 has the Fourier coefficients c_n. Of course, $c_{-n} = \bar{c}_n$, since u is real-valued. So we obtain u_r from $u(e^{i\theta})$ by multiplying the Fourier coefficient c_n by $r^{|n|}$. This means only that u_r is the convolution of $u(e^{i\theta})$ with the function P_r, whose Fourier coefficients are $r^{|n|}$:

$$P_r(\theta) = \sum_{n=-\infty}^{\infty} r^{|n|} e^{in\theta}$$
$$= C_r(\theta) + \overline{C_r(\theta)} - 1$$
$$= 2\,\mathrm{Re}\,C_r(\theta) - 1$$
$$= \mathrm{Re}\,[2C_r(\theta) - 1]$$
$$= \mathrm{Re}\left[\frac{1 + re^{i\theta}}{1 - re^{i\theta}}\right]$$
$$= \frac{1 - r^2}{1 - 2r\cos\theta + r^2}.$$

This family of functions P_r is called **Poisson's kernel**. We have just noted that for any real harmonic function in the closed disc we have

$$u_r(\theta) = u(re^{i\theta}) = \frac{1}{2\pi}\int_{-\pi}^{\pi} u(t)P_r(\theta - t)dt$$

where, as usual, $u(t)$ denotes $u(e^{it})$. Of course, it immediately follows that this Poisson integral formula holds for any complex-valued harmonic function in the closed disc. In particular, it holds for an analytic function f. Thus, both the Cauchy kernel C_r and the Poisson kernel P_r reproduce analytic functions from their boundary values by convolution. It is easy to see why this is so. The functions P_r and C_r have the same Fourier coefficients on the non-negative integers. Consequently, when we convolve them with an "analytic" function on the circle, the results are the same. The difference is that whereas the Fourier coefficients of P_r are symmetric about zero on the integers:

$$\frac{1}{2\pi}\int_{-\pi}^{\pi} e^{-in\theta}P_r(\theta)d\theta = r^{|n|},$$

the Fourier coefficients of C_r vanish on the negative integers:

$$\frac{1}{2\pi}\int_{-\pi}^{\pi} e^{-in\theta}C_r(\theta)d\theta = \begin{cases} r^n, & n \geq 0 \\ 0, & n < 0. \end{cases}$$

This vanishing of the negative Fourier coefficients of C_r simply means that C_r is "orthogonal" to the conjugate of any analytic function which vanishes at the origin; i.e., if f is analytic in the closed disc, then

$$\overline{f(e^{i\theta})} = \sum_{n=0}^{\infty} \bar{a}_n e^{-in\theta}$$

and thus

$$\frac{1}{2\pi}\int_{-\pi}^{\pi} \overline{f(e^{it})}C_r(\theta - t)dt = \bar{a}_0 = \overline{f(0)}.$$

On the other hand,

$$\frac{1}{2\pi}\int_{-\pi}^{\pi}\overline{f(e^{it})}P_r(\theta-t)dt = \overline{f(re^{i\theta})}$$

because $\bar{f}$ is harmonic. The kernel

$$H_r(\theta) = 2C_r(\theta) - 1$$
$$= \frac{1+re^{i\theta}}{1-re^{i\theta}}$$

is also interesting, in part because

$$P_r(\theta) = \operatorname{Re} H_r(\theta).$$

But it is of more interest, because if $f = u + iv$ is analytic in the closed disc, *and if $f(0)$ is real*, then

$$f(re^{i\theta}) = \frac{1}{2\pi}\int_{-\pi}^{\pi} u(t)H_r(\theta-t)dt,$$

i.e., $f_r = u*H_r$. This is easy to see because $u = \frac{1}{2}(f+\bar{f})$ and so

$$\frac{1}{2\pi}\int_{-\pi}^{\pi}\tfrac{1}{2}(f(e^{it})+\overline{f(e^{it})})H_r(\theta-t)dt$$

$$= \frac{1}{2\pi}\int_{-\pi}^{\pi}\tfrac{1}{2}(f+\bar{f})\cdot[2C_r(\theta-t)-1]dt$$

$$= \frac{1}{2\pi}\int_{-\pi}^{\pi}f(e^{it})C_r(\theta-t)dt + \frac{1}{2\pi}\int_{-\pi}^{\pi}\overline{f(e^{it})}C_r(\theta-t)dt$$

$$\quad - \frac{1}{4\pi}\int_{-\pi}^{\pi}[f(e^{it})+\overline{f(e^{it})}]dt$$

$$= f(re^{i\theta}) + \overline{f(0)} - \operatorname{Re} f(0).$$

So, if $f(0)$ is real, $f_r = u*H_r$. This formula can be rewritten

$$f(re^{i\theta}) = \frac{1}{2\pi}\int_{-\pi}^{\pi} u(e^{it})\frac{e^{it}+re^{i\theta}}{e^{it}-re^{i\theta}}dt$$

or

$$f(z) = \frac{1}{2\pi}\int_{-\pi}^{\pi} u(e^{it})\frac{e^{it}+z}{e^{it}-z}dt.$$

Suppose we let

$$Q_r = \operatorname{Im} H_r.$$

This kernel is called the **conjugate Poisson kernel**. We see from above that

$$v(re^{i\theta}) = \frac{1}{2\pi}\int_{-\pi}^{\pi} u(t)Q_r(\theta-t)dt$$

produces the harmonic conjugate v of u, which vanishes at the origin. Of course, $P_r(\theta)$ and $Q_r(\theta)$ are conjugate harmonic functions in the disc.

Boundary Values

To begin seriously our discussion of the boundary behavior of harmonic functions, we shall consider the problem of starting with a function on the unit circle and extending it to a harmonic function in the disc. The original problem of this sort was the Dirichlet problem: given a real-valued continuous function f on the unit circle, find a continuous function on the closed disc which agrees with f on the circle and which is harmonic in the open disc D. This problem is completely solved by the Poisson integral formula. All one needs to show is that the family of functions P_r, $0 \leq r < 1$, is an approximate identity for L^1 of the circle. Since

$$P_r(\theta) = \frac{1 - r^2}{1 - 2r\cos\theta + r^2}$$

we see the following:

(i) $P_r(\theta) \geq 0$ (and P_r is continuous on the circle);

(ii) $\dfrac{1}{2\pi}\displaystyle\int_{-\pi}^{\pi} P_r(\theta)d\theta = 1, \quad 0 \leq r < 1$

(because the above integral is the value of the constant function 1 at $z = r$);

(iii) if $0 < \delta < \pi$, then

$$\limsup_{r \to 1} {}_{|\theta| \geq \delta} |P_r(\theta)| = 0.$$

For if $\delta \leq |\theta| \leq \pi$, then $P_r(\theta) \leq \dfrac{1 - r^2}{1 - 2r\cos\delta + r^2}$.

Theorem. *Let* f *be a complex-valued function in* L^p *of the unit circle, where* $1 \leq p < \infty$. *Define* f *in the unit disc by*

$$f(re^{i\theta}) = \frac{1}{2\pi}\int_{-\pi}^{\pi} f(t) P_r(\theta - t) dt.$$

Then the extended function f *is harmonic in the open unit disc, and, as* $r \to 1$, *the functions* $f_r(\theta) = f(re^{i\theta})$ *converge to* f *in the* L^p-*norm. If* f *is continuous on the unit circle, the* f_r *converge uniformly to* f; *thus, the extended* f *is continuous on the closed disc, harmonic in the interior.*

Proof. Since $\{P_r\}$ is an approximate identity, the proofs of the L^p convergence and the uniform convergence are just the same as the corresponding proofs about Cesaro means. The only thing which is basically different here is that we also regard the family of functions $\{f_r\}$ as a harmonic function on the open disc. There are various ways of seeing why this function is harmonic. One way is to observe that if the original f is real-valued then the function $f(re^{i\theta})$ is the real part of the analytic function

$$g(z) = \frac{1}{2\pi} \int_{-\pi}^{\pi} f(t) \frac{e^{it} + z}{e^{it} - z} dt.$$

Theorem. *Let* f *be a bounded Baire function on the unit circle and*

$$f(re^{i\theta}) = \frac{1}{2\pi} \int_{-\pi}^{\pi} f(t) P_r(\theta - t) dt.$$

The extended f *is a bounded harmonic function in the open disc and, as* $r \to 1$, *the functions* $f_r(\theta) = f(re^{i\theta})$ *converge to* f *in the weak-star topology on* L^{∞}.

Proof. $\{P_r\}$ is an approximate identity.

Theorem. *Let* μ *be a finite complex Baire measure on the unit circle and let*

$$f(re^{i\theta}) = \int_C P_r(\theta - t) d\mu(t).$$

Then f *is harmonic in the open disc and the measures*

$$d\mu_r = \frac{1}{2\pi} f_r(\theta) d\theta$$

converge to μ *in the weak-star topology on measures.*

In all the above cases, it would seem convenient to say that the harmonic function $f(re^{i\theta})$ is the **Poisson integral** of the corresponding function or measure on the circle. This will save us some words as we proceed to reverse the process. Just as we did for Cesaro means, we will now ask: given a harmonic function in the disc, how do we ascertain if it is the Poisson integral of some type of function or measure on the unit circle? If f is harmonic, then

$$f(re^{i\theta}) = \sum_{n=-\infty}^{\infty} c_n r^{|n|} e^{in\theta}$$

so the question is actually: when is $\{c_n\}$ the sequence of Fourier coefficients of some type of function or measure? Of course, the answer will read just as it did for Cesaro means, and so will the proof.

Theorem. *Let* f *be a complex-valued harmonic function in the open unit disc, and write*

$$f_r(\theta) = f(re^{i\theta}).$$

(i) *If* $1 < p \leq \infty$, *then* f *is the Poisson integral of an* L^p *function on the unit circle if and only if the functions* f_r *are bounded in* L^p-*norm*.

(ii) f *is the Poisson integral of an integrable function on the circle if and only if the* f_r *converge in the* L^1-*norm*.

(iii) f *is the Poisson integral of a continuous function on the unit circle if and only if the* f_r *converge uniformly.*

(iv) f *is the Poisson integral of a finite complex Baire measure on the circle if and only if the* f_r *are bounded in* L^1-*norm*.

(v) f *is the Poisson integral of a finite positive Baire measure if and only if* f *is non-negative.*

We should make a few comments about the various parts of this theorem. The L^∞ part of (i) is often called Fatou's theorem. The interesting part of it is the fact that any bounded harmonic function in the disc is the Poisson integral of a bounded Baire function on the circle. Part (v) is often called Herglotz's theorem: every non-negative harmonic function is the Poisson integral of a positive measure. One should note that in any of the cases above the harmonic function f is real-valued if and only if the corresponding L^p function or measure is real.

Fatou's Theorem

The results we have obtained so far about harmonic functions are completely analogous to our former results about Cesaro means. Indeed, one can view these results simply as another way of summing Fourier series (Abel summability). One theorem on Cesaro summability which we did not prove is Lebesgue's theorem: if f is an integrable function on $[-\pi, \pi]$, the Cesaro means of the Fourier series for f converge to f pointwise almost everywhere. This result has its analogue in Abel summability: if we extend f to a harmonic function in the unit disc, the functions f_r converge pointwise to f almost everywhere. This is a theorem of Fatou, which we shall now prove.

Theorem (Fatou). *Let μ be a finite complex Baire measure on the unit circle, and let* f *be the harmonic function in the unit disc defined by*

$$f(r, \theta) = \int P_r(\theta - t) d\mu(t).$$

Let θ_0 be any point where μ is differentiable with respect to Lebesgue measure. Then

$$\lim_{r \to 1} f(r, \theta_0) = 2\pi \left(\frac{d\mu}{d\theta}\right)(\theta_0) = 2\pi\mu'(\theta_0).$$

In fact,

$$\lim f(r, \theta) = 2\pi\mu'(\theta_0)$$

as the point $z = re^{i\theta}$ *approaches* $e^{i\theta_0}$ *along any path in the open disc which is not tangent to the unit circle.*

Proof. The measure μ is induced by a complex-valued function F, of bounded variation on the interval $[-\pi, \pi]$:

$$\int g d\mu = \int g dF.$$

The theorem states that if F is differentiable at θ_0, then

$$\lim f(r, \theta) = 2\pi F'(\theta_0)$$

as $re^{i\theta} \to e^{i\theta_0}$ along any non-tangential path, i.e., $f(z)$ approaches the derivative of F (or μ) with respect to *normalized* Lebesgue measure. For the proof, let us first observe that the theorem is trivially true for $d\mu = d\theta$. So, without loss of generality, we may (by subtracting a constant multiple of $d\theta$ from $d\mu$) assume that $\mu(C) = 0$. Then F will satisfy $F(-\pi) = F(\pi)$. Now let

$$F(r, \theta) = \frac{1}{2\pi} \int_{-\pi}^{\pi} P_r(\theta - t) F(t) dt.$$

Then

$$\frac{1}{2\pi} \int_{-\pi}^{\pi} P_r'(\theta - t) F(t) dt = -\frac{1}{2\pi} P_r(\theta - t) F(t) \Big|_{-\pi}^{\pi} + \frac{1}{2\pi} \int_{-\pi}^{\pi} P_r(\theta - t) dF(t).$$

So

$$\frac{1}{2\pi} f(r, \theta) = \frac{1}{2\pi} \int_{-\pi}^{\pi} P_r'(\theta - t) F(t) dt.$$

We first prove the radial convergence, since its proof is neater.

$$f(r, \theta) = \int_{-\pi}^{\pi} P_r'(t) F(\theta - t) dt$$

$$= \int_{0}^{\pi} + \int_{-\pi}^{0}$$

$$= \int_{0}^{\pi} P_r'(t) [F(\theta - t) - F(\theta + t)] dt$$

$$= \int_{0}^{\pi} [-\sin t P_r'(t)] \frac{F(\theta + t) - F(\theta - t)}{\sin t} dt.$$

Since P_r' is an odd function, we have

$$\frac{1}{2\pi} f(r, \theta) = \frac{r}{2\pi} \int_{-\pi}^{\pi} K_r(t) \frac{F(\theta + t) - F(\theta - t)}{2 \sin t} dt$$

where

$$K_r(t) = -\frac{1}{r} \sin t P_r'(t).$$

Now it is easy to verify that $\{K_r\}$, $0 < r < 1$, is an approximate identity for L^1. If F is differentiable at θ_0, then the function

$$G(t) = \frac{F(\theta_0 + t) - F(\theta_0 - t)}{2 \sin t}$$

is continuous at $t = 0$ with the value

$$G(0) = \lim_{t \to 0} \frac{F(\theta_0 + t) - F(\theta_0 - t)}{2 \sin t} = F'(\theta_0).$$

Since $\{K_r\}$ is an approximate identity, and since G is continuous at 0,

$$\lim_{r \to 1} \frac{1}{2\pi} f(r, \theta_0) = \lim_{r \to 1} \frac{1}{2\pi} \int_{-\pi}^{\pi} K_r(t) G(t) dt$$
$$= G(0)$$
$$= F'(\theta_0).$$

Now for the non-tangential convergence. Suppose we have an arc in the disc which approaches $e^{i\theta_0}$ non-tangentially. This means that we have two continuous functions $r = r(\alpha)$, $\theta = \theta(\alpha)$ defined for $0 \leq \alpha \leq 1$ such that $0 \leq r(\alpha) < 1$ for $0 \leq \alpha < 1$, $r(1) = 1$, $\theta(1) = \theta_0$. It is no loss of generality to assume that $\theta_0 = 0$. The non-tangential nature of the arc then means that

$$\frac{\theta(\alpha)}{1 - r(\alpha)}$$

is bounded for $\alpha < 1$. Let

$$K_\alpha(t) = \sin t P_r'(\theta - t) \quad \begin{cases} \theta = \theta(\alpha) \\ r = r(\alpha) \end{cases}$$

so that

$$\frac{1}{2\pi} f(r(\alpha), \theta(\alpha)) = \frac{1}{2\pi} \int_{-\pi}^{\pi} K_\alpha(t) \frac{F(t)}{\sin t} dt.$$

Now we may also assume that $F(0) = 0$, for this can be arranged by subtracting a constant from F, which will change neither dF nor the condition $F(-\pi) = F(\pi)$. Then

$$F'(0) = \lim_{t \to 0} \frac{F(t)}{\sin t}.$$

Now we shall prove that the functions K_α satisfy these conditions:

(i) $\int_{-\pi}^{\pi} |K_\alpha(t)| dt$ is bounded as $\alpha \to 1$;

(ii) $\lim_{\alpha \to 1} \frac{1}{2\pi} \int_{-\pi}^{\pi} K_\alpha(t) dt = 1$;

(iii) If $0 < \delta < \pi$, then
$$\lim_{\alpha \to 1} \sup_{\delta \leq |t| \leq \pi} |K_\alpha(t)| = 0.$$

Condition (iii) is easy to verify, since

$$\sin t P_r'(\theta - t) = 2r(1 - r^2) \cdot \frac{\sin t \sin(t - \theta)}{[1 - 2r \cos(\theta - t) + r^2]^2}$$
$$= \frac{2r \sin t \sin(t - \theta)}{1 - 2r \cos(\theta - t) + r^2} P_r(\theta - t)$$

so that if $|t| \geq \delta$ while θ is near zero, $\sin t P_r'(\theta - t)$ is small for r near 1. Condition (ii) follows from

$$\frac{1}{2\pi} \int_{-\pi}^{\pi} \sin t P_r'(\theta - t) dt = \frac{1}{2\pi} \int_{-\pi}^{\pi} \cos t P_r(\theta - t) dt = r \cos \theta.$$

It is in verifying condition (i) that we use the non-tangential nature of the arc:

$$\frac{1}{2\pi}\int_{-\pi}^{\pi}|K_\alpha(t)|dt = \frac{1}{2\pi}\int_{-\pi}^{\pi}|\sin(\theta+t)P_r'(t)|dt$$

$$\leq |\sin\theta|\frac{1}{2\pi}\int_{-\pi}^{\pi}|P_r'(t)|dt + \frac{1}{2\pi}\int_{-\pi}^{\pi}|\sin t P_r'(t)|dt.$$

Since $-\frac{1}{r}\sin t P_r'(t)$ is an approximate identity, the second integral is bounded as $r \to 1$. Also,

$$|\sin\theta|\cdot\frac{1}{2\pi}\int_{-\pi}^{\pi}|P_r'(t)|dt = |\sin\theta|\cdot\frac{1}{\pi}\int_{-\pi}^{0}|P_r'(t)|dt$$

$$= |\sin\theta|\frac{1}{\pi}\int_{-\pi}^{0}P_r'(t)dt$$

$$= \frac{1}{\pi}|\sin\theta|[P_r(0) - P_r(-\pi)]$$

$$= \frac{1}{\pi}|\sin\theta|\left[\frac{1+r}{1-r} - \frac{1-r}{1+r}\right]$$

$$\leq \frac{1+r}{\pi}\frac{|\theta|}{1-r}.$$

Since $|\theta|/(1-r)$ is bounded on our arc, we conclude that $\int|K_\alpha|$ is bounded as $\alpha \to 1$.

With these three properties of the K_α, we finish the proof. Put

$$G(t) = \frac{F(t)}{\sin t} - F'(0) \quad \text{and} \quad I(\alpha) = \frac{1}{2\pi}\int K_\alpha.$$

Then

$$\frac{1}{2\pi}f(r(\alpha),\theta(\alpha)) - I(\alpha)F'(0) = \frac{1}{2\pi}\int_{-\pi}^{\pi}G(t)K_\alpha(t)dt.$$

Since G is continuous at 0 with $G(0) = 0$, for δ small the integral

$$\int_{-\delta}^{\delta}GK_\alpha$$

is small by condition (i) on the K_α. Then

$$\int_{|t|\geq\delta}GK_\alpha$$

is small by condition (iii). Since $\lim_{\alpha\to 1}I(\alpha) = 1$, we see that

$$\lim_{\alpha\to 1}\frac{1}{2\pi}f(r(\alpha),\theta(\alpha)) = F'(0).$$

This completes the proof.

Corollary. *Let* f *be a Lebesgue-integrable function on the unit circle. Then the Poisson integral of* f *has a non-tangential limit at almost every point of the unit circle, and these limits are almost everywhere equal to* f. *More generally, the Poisson integral of a finite measure* μ *has non-tangential limits equal almost everywhere to the derivative of* μ *with respect to normalized Lebesgue measure.*

Proof. Let μ be a finite (complex) Baire measure on the circle and let $d\mu = \frac{1}{2\pi} f d\theta + d\mu_s$ be the Lebesgue decomposition for μ. Then μ is differentiable almost everywhere and $\frac{d\mu}{d\theta} = \frac{1}{2\pi} f$ almost everywhere. Now apply Fatou's theorem.

Corollary. *Let* f *be a complex-valued harmonic function in the unit disc and suppose that the integrals*

$$\int_{-\pi}^{\pi} |f(re^{i\theta})|^p d\theta$$

are bounded as r $\to$ 1 *for some* p, $1 \leq p < \infty$. *Then for almost every* θ *the radial limits*

$$\tilde{f}(\theta) = \lim_{r \to 1} f(re^{i\theta})$$

exist and define a function $\tilde{f}$ *in* L^p *of the circle. If* p > 1 *then* f *is the Poisson integral of* $\tilde{f}$. *If* p = 1 *then* f *is the Poisson integral of a (unique) finite measure whose absolutely continuous part is* $\frac{1}{2\pi} \tilde{f} d\theta$. *If* f *is a bounded harmonic function, the boundary values exist almost everywhere and define a bounded measurable function* $\tilde{f}$ *whose Poisson integral is* f.

Of course, the limits in the last Corollary exist non-tangentially as well as radially. For emphasis we might also state the following.

Corollary. *A non-negative harmonic function in the unit disc has non-tangential limits at almost every point of the unit circle.*

One conclusion from the various theorems above is the following. Suppose $1 \leq p \leq \infty$ and we consider the class of harmonic functions f in the open disc such that the functions $f_r(\theta) = f(re^{i\theta})$ are bounded in L^p-norm. This class of harmonic functions forms a Banach space under the norm

$$||f|| = \lim_{r \to 1} ||f_r||_p.$$

For $1 < p \leq \infty$ this Banach space is isomorphic to L^p of the unit circle. The isomorphism is $f \to \tilde{f}$, where $\tilde{f}$ is the boundary function for f. If $1 < p < \infty$, we have not only

$$||\tilde{f}||_p = \lim_{r \to 1} ||f_r||_p$$

but also $\lim_{r \to 1} ||\tilde{f} - f_r||_p = 0$. For $p = 1$ this Banach space is isomorphic to the space of finite (Baire) measures on the circle, the isomorphism being $f \to \mu$, where f is the Poisson integral of μ.

H^p Spaces

Our results about harmonic functions apply in particular to analytic functions. If $0 < p \leq \infty$ we denote by H^p (H for Hardy) the class of analytic functions f in the unit disc for which the functions $f_r(\theta) = f(re^{i\theta})$ are bounded in L^p-norm as $r \to 1$. If $1 \leq p \leq \infty$, then H^p is a Banach space under the norm

$$||f|| = \lim_{r \to 1} ||f_r||_p$$

i.e., H^p is a closed subspace of the corresponding space of harmonic functions. If $1 < p \leq \infty$, we can then identify H^p with a closed subspace of L^p of the circle. This space we shall also denote by H^p, because of the isomorphism. It consists of all functions f in L^p whose Poisson integrals are analytic on the disc, i.e., all f in L^p such that

$$\int_{-\pi}^{\pi} f(\theta) e^{in\theta} d\theta = 0, \quad n = 1, 2, 3, \ldots.$$

When $p = 1$ we obtain an identification of H^1 with the closed space of finite measures μ on the circle which are "analytic":

$$\int_{-\pi}^{\pi} e^{in\theta} d\mu(\theta) = 0, \quad n = 1, 2, 3, \ldots.$$

Now it is here that a very significant difference occurs between the harmonic and analytic cases. A theorem of F. and M. Riesz states that any measure μ which is analytic as above is necessarily absolutely continuous with respect to Lebesgue measure. This theorem makes it possible for us to identify H^1 with the space of Lebesgue-integrable functions on the circle such that

$$\int_{-\pi}^{\pi} e^{in\theta} f(\theta) d\theta = 0, \quad n = 1, 2, 3, \ldots.$$

Thus, our next task will be the proof of the theorem of F. and M. Riesz, along with some related results.

NOTES

Fundamental facts about analytic and harmonic functions can be found in Titchmarsh [87], Ahlfors [2] or many other places. For more about the Dirichlet problem, see Courant's book [21]. Fatou's paper is [27]. The various boundary value results here can be found in the books by Zygmund [98], Evans [24], Bieberbach [9], Nevanlinna [64], and Privaloff [70]. For analytic functions of class H^p see F. Riesz [71] and F. and M. Riesz [72].

EXERCISES

1. If f is harmonic and $zf(z)$ is harmonic, then f is analytic.

2. If $\{c_n\}$ is a bounded sequence, then
$$f(r, \theta) = \sum_{n=-\infty}^{\infty} c_n r^{|n|} e^{in\theta}$$
is harmonic in the disc.

3. (a) Verify that $P_r(\theta)$ is harmonic.
(b) If u is harmonic and $0 \leq u(r, \theta) \leq P_r(\theta)$, then $u(r, \theta) = \lambda P_r(\theta)$ for some constant λ.
(c) The set of all non-negative harmonic functions in the unit disc which have the value 1 at the origin is a convex set of functions. What are the extreme points of this set?

4. Let f be an analytic function in the unit disc *without zeros* satisfying $|f| \leq 1$. Prove that
$$\sup_{|z| \leq 1/5} |f(z)|^2 \leq \inf_{|z| \leq 1/7} |f(z)|.$$

5. For a real-valued harmonic function u in the disc, let v denote the harmonic conjugate vanishing at the origin. Is the map from u_r to v_r continuous in the sup norm? The L^2-norm?

6. For a real-valued harmonic function u, the following are equivalent: (i) u is the difference of two non-negative harmonic functions; (ii) the L^1 norms of the functions $u_r(\theta) = u(r, \theta)$ are bounded.

7. Give an example of an analytic function in the unit disc which is in no class H^p but which has non-tangential limits at almost every point of the unit circle.

8. Give an example of an analytic function in the disc which does not have non-tangential limits at almost every point of the unit circle.

9. Prove that the set of all analytic functions in the unit disc for which
$$\iint_D |f(z)|^2 dx dy < \infty$$
is a Hilbert space, using the square-root of the above integral as norm. Prove that H^2 is a (linear) subspace of this Hilbert space. Is H^2 a dense subspace? Find an orthonormal basis for this Hilbert space.

10. Prove Herglotz's theorem: Every analytic function in the unit disc with values in the right half-plane such that $f(0) > 0$ has the form
$$f(z) = \int \frac{e^{i\theta} + z}{e^{i\theta} - z} d\mu(\theta)$$
where μ is a finite positive measure on the unit circle.

The set of all f analytic in $|z| < 1$ satisfying $\operatorname{Re} f \geq 0$ and $f(0) = 1$ is a convex set of functions. What are the extreme points of this convex set?

11. Let u be the real harmonic function in the disc determined by some finite real measure on the circle. Prove that the associated conjugate harmonic functions v have non-tangential limits at almost every point of the circle.

12. Show that Fatou's theorem extends to the following situation. If f is a real-valued integrable function on the unit circle, and if θ_0 is a point such that $\lim_{\theta \to \theta_0} f(\theta) = +\infty$, then
$$\lim_{r \to 1} f(r, \theta_0) = +\infty.$$

CHAPTER 4

THE SPACE H^1

The Helson-Lowdenslager Approach

Most of the theorems in this chapter generalize to the context of a certain class of function algebras known as Dirichlet algebras. We shall give the proofs for the case of the unit disc, but in such a manner that they generalize readily. We shall describe the generalizations later. It may be that these proofs are not always the shortest possible as applied to the classical case; however, they have an undeniable elegance. These first few proofs are due to Helson and Lowdenslager.

We denote by A the collection of functions which are continuous on the closed unit disc and analytic at each interior point. Then A is a uniformly closed linear algebra of continuous complex-valued functions on the closed disc. In particular, A is Banach space under the sup norm

$$||f||_\infty = \sup_{|z| \leq 1} |f(z)|.$$

Each f in A is (of course) the Poisson integral of its boundary values:

$$f(re^{i\theta}) = \frac{1}{2\pi} \int_{-\pi}^{\pi} f(e^{it}) P_r(\theta - t) dt.$$

Also

$$||f||_\infty = \sup |f(e^{it})|.$$

This is easily seen from the maximum modulus principle for analytic functions. Or, if one wishes, one may deduce this fact (i.e., the maximum modulus principle for functions in A) directly from the Poisson formula. Thus, we may identify the functions in A with their boundary values, obtaining an isomorphism between A and the Banach space of continuous functions on the circle such that

$$\int_{-\pi}^{\pi} f(\theta) e^{in\theta} d\theta = 0, \quad n = 1, 2, 3, \ldots.$$

This algebra of continuous functions we shall also denote by A. The trigonometric polynomials of the form

$$P(\theta) = \sum_{k=0}^{n} a_k e^{ik\theta}$$

are in A and are uniformly dense in A. This follows, for example, from the fact that if f is continuous on the circle and if the Fourier coefficients of f vanish on the negative integers, then the Cesaro means of the Fourier series for f comprise a sequence of trigonometric polynomials of the above form which converge uniformly to f. This corresponds to the fact that A, as an algebra on the disc, consists of all functions which are uniformly approximable by polynomials in z:

$$P(z) = \sum_{k=0}^{n} a_k z^k.$$

The property of A (on the circle) which we want is Fejer's theorem, in the following form.

Theorem. *The real parts of the functions in* A *are uniformly dense in the space of real-valued continuous functions on the unit circle. In other words, if μ is a finite real Baire measure on the circle such that $\int fd\mu = 0$ for every* f *in* A, *then μ is the zero measure.*

Proof. The real parts of the functions in A include every trigonometric polynomial of the form

$$P(\theta) = \sum_{k=-n}^{n} c_k e^{ik\theta}, \quad c_{-k} = \bar{c}_k$$

that is, every real-valued trigonometric polynomial. If f is real-valued and continuous on the circle, every Cesaro mean for f is such a polynomial. Hence, such polynomials are dense in the real continuous functions.

If μ is a finite real measure on the circle which is "orthogonal" to every f in A, then μ is orthogonal to the real part of every f in A. So μ is orthogonal to every real continuous function and must be the zero measure.

Corollary. *If μ is a finite real measure on the circle such that $\int fd\mu = 0$ for every* f *in* A *which vanishes at the origin, then μ is a constant multiple of Lebesgue measure.*

Proof. Let

$$\lambda = \int d\mu \quad \text{and} \quad d\mu_1 = d\mu - \frac{1}{2\pi}\lambda d\theta.$$

Then μ_1 is a real measure which is orthogonal to every f in A:

$$\int fd\mu_1 = \int [f - f(0)]d\mu_1 + f(0) \int d\mu_1 = 0 + 0 = 0.$$

Thus

$$\mu_1 = 0 \quad \text{or} \quad d\mu = \frac{1}{2\pi}\lambda d\theta.$$

We shall be working for some time entirely on the unit circle. Thus

A, H^2, etc. will be spaces of functions on the unit circle. One of the theorems we shall prove is the theorem of F. and M. Riesz which we mentioned earlier: an "analytic" measure on the unit circle is absolutely continuous with respect to Lebesgue measure. We shall also prove Szegö's theorem. The setting for this theorem is as follows. We are given a finite *positive* measure μ on the circle, and we wish to know in the Hilbert space $L^2(d\mu)$ what the distance is from the constant function 1 to the subspace spanned by the functions in A which "vanish at the origin." That is, we wish to compute

$$\inf \int |1 - f|^2 d\mu, \quad f \text{ in } A \text{ and } \int f d\theta = 0.$$

We shall be particularly interested in characterizing the measures μ for which the infimum is zero, i.e., the measures for which 1 is in the closed subspace of $L^2(d\mu)$ spanned by the functions $e^{in\theta}$, $n \geq 1$. Szegö's theorem states that the square-distance (infimum) above is equal to

$$\exp\left[\frac{1}{2\pi} \int_{-\pi}^{\pi} \log h(\theta) d\theta\right]$$

where h is the derivative of μ with respect to normalized Lebesgue measure:

$$d\mu = \frac{1}{2\pi} h d\theta + d\mu_s, \quad \mu_s \text{ singular}.$$

In particular, this distance depends only upon the absolutely continuous part of μ. Actually, Szegö proved the distance formula for absolutely continuous measures, and Kolmogoroff and Krein extended it to the general case.

Our program will be this. We begin to look at Szegö's theorem and obtain a preliminary result. This preliminary result can be used to prove the Riesz theorem. Then we return and complete the proof of Szegö's theorem.

Let A_0 denote the set of functions f in A for which $\int f d\theta = 0$. If μ is a positive measure, we are interested in the $L^2(d\mu)$ distance from 1 to A_0. The square of this distance will be

$$\inf_{f \in A_0} \int |1 - f|^2 d\mu = \int |1 - F|^2 d\mu$$

where F is the orthogonal projection of 1 into the closed subspace of $L^2(d\mu)$ which is spanned by the functions in A_0.

Theorem. *Let μ be a finite positive Baire measure on the circle and suppose 1 is not in the closed subspace of $L^2(d\mu)$ which is spanned by the functions in A_0. Let F be the orthogonal projection of 1 into that closed subspace.*

(i) *The measure $|1 - F|^2 d\mu$ is a non-zero constant multiple of Lebesgue measure. In particular, Lebesgue measure is absolutely continuous with respect to μ.*

(ii) *The function* $(1-F)^{-1}$ *is in* H^2.
(iii) *If* h *is the derivative of* μ *with respect to normalized Lebesgue measure, then the function* $(1-F)h$ *is in*

$$L^2 = L^2\left(\frac{1}{2\pi}d\theta\right).$$

Proof. Let S be the closed subspace of $L^2(d\mu)$ spanned by A_0. Since F is the orthogonal projection of 1 into S, the function $(1-F)$ is orthogonal to S. That is, $(1-F)$ is orthogonal to A_0. But $(1-F)$ is also orthogonal to $(1-F)f$ for every f in A_0, because F is the limit in $L^2(d\mu)$ of a sequence of elements f_n in A_0, and if f is a fixed element of A_0, then $f(1-f_n)$ is in A_0 and converges to $f(1-F)$. The statement that $(1-F)$ is orthogonal to $(1-F)f$ for all f in A_0 says

$$\int f|1-F|^2 d\mu = 0, \quad f \text{ in } A_0.$$

Hence the measure $|1-F|^2 d\mu$ is a constant multiple of Lebesgue measure. That constant is not zero because $1 - F \neq 0$, i.e., 1 is not in S. Thus (i) is proved.

To prove (ii), observe that

$$|1-F|^2 d\mu = k d\theta, \quad k \neq 0$$

and so, if μ_a denotes the absolutely continuous part of μ,

$$d\mu_a = |1-F|^{-2} k d\theta$$

proving that $(1-F)^{-1}$ is in $L^2 = L^2\left(\frac{1}{2\pi}d\theta\right)$. Suppose f is in A_0. Then

$$k \int (1-F)^{-1} f d\theta = k \int (1-\bar{F})f|1-F|^{-2} d\theta$$
$$= \int (1-\bar{F})f d\mu$$
$$= 0$$

because $(1-F)$ is orthogonal to f in $L^2(d\mu)$. Since this holds for every f in A_0 (in particular for $f(\theta) = e^{in\theta}$, $n = 1, 2, 3, \ldots$) we see that $(1-F)^{-1}$ is in H^2.

To prove (iii), suppose $d\mu = \frac{1}{2\pi} h d\theta + d\mu_s$, μ_s singular. Since $|1-F|^2 d\mu$ is a constant multiple of $d\theta$, the function $(1-F)$ vanishes almost everywhere with respect to μ_s. Thus

$$|1-F|^2 d\mu = \frac{1}{2\pi}|1-F|^2 h d\theta.$$

But $|1-F|^2 d\mu = k d\theta$, so $|1-F|^2 h$ is equal to a non-zero constant ρ, almost everywhere with respect to $d\theta$. But then

$$|1-F|h = \rho|1-F|^{-1}$$

and since $(1-F)^{-1}$ is in $L^2\left(\frac{1}{2\pi}d\theta\right)$, so is $(1-F)h$.

Corollary 1. *If μ is a positive measure on the circle with absolutely continuous part μ_a, then*
$$\inf_{f \in A_0} \int |1-f|^2 d\mu = \inf_{f \in A_0} \int |1-f|^2 d\mu_a.$$
In particular, for any singular μ the function 1 is in the $L^2(d\mu)$ closure of A_0.

Proof. Let F be the orthogonal projection of 1 into the $L^2(d\mu)$ closed span of A_0. The square of the distance from 1 to A_0 in $L^2(d\mu)$ is then $\int |1-F|^2 d\mu$. As we noted in the theorem, the function $(1-F)$ vanishes almost everywhere for the singular part of μ. If we regard F as an element of $L^2(d\mu_a)$, then F is in the closure of A_0 in that space. Furthermore, since $(1-F)$ vanishes almost everywhere with respect to $d\mu_s$, it is easily seen that $(1-F)$ is orthogonal to A_0 in $L^2(d\mu_a)$. Thus $(1-F)$ is also the minimizing function for μ_a, and the proof is complete.

Corollary 2. *Let μ be a finite complex Baire measure on the circle which is orthogonal to A_0, i.e., $\int f d\mu = 0$ for all f in A_0. Then the absolutely continuous and singular parts of μ are separately orthogonal to A_0.*

Proof. Let ρ be any finite *positive* measure with these two properties:

(i) μ is absolutely continuous with respect to ρ and the Radon-Nikodym derivative $d\mu/d\rho$ is bounded.

(ii) $\dfrac{d\rho}{d\theta} \geq \dfrac{1}{2\pi}$.

If $d\mu = \dfrac{1}{2\pi} h d\theta + d\mu_s$, then one such measure ρ is
$$d\rho = \frac{1}{2\pi}(1+|h|)d\theta + d|\mu_s|$$
where $d|\mu_s|$ is the total variation of the complex measure μ_s. Or in place of $d|\mu_s|$ use the sum of the total variations of the real and imaginary parts of μ_s.

Let f be in A_0. By property (ii) of ρ
$$\int |1-f|^2 d\rho \geq \frac{1}{2\pi}\int |1-f|^2 d\theta \geq 1.$$

If F is the orthogonal projection of 1 into the closed subspace of $L^2(d\rho)$ spanned by A_0, then
$$\int |1-F|^2 d\rho \geq 1.$$
By the theorem above, $(1-F)^{-1}$ is in H^2 and $(1-F)(1+|h|)$ is in
$$L^2 = L^2\left(\frac{1}{2\pi}d\theta\right).$$

So $(1-F)h$ is in L^2.

Suppose g is in A_0. Then
$$\int (1 - F)g\,d\mu = 0.$$
For, choose a sequence of elements f_n in A_0 which converge to F in $L^2(d\rho)$. Since $d\mu/d\rho$ is bounded, we have
$$\int (1 - F)g\,d\mu = \lim_{n\to\infty} \int (1 - f_n)g\,d\mu.$$
Each $(1 - f_n)g$ is in A_0, and since μ is orthogonal to A_0, we have
$$\int (1 - F)g\,d\mu = 0, \quad g \text{ in } A_0.$$
Now $(1 - F)$ vanishes almost everywhere with respect to the singular part of ρ, hence, vanishes almost everywhere $d\mu_s$. Thus
$$(1 - F)d\mu = \frac{1}{2\pi}(1 - F)h\,d\theta$$
so we have
$$\int (1 - F)gh\,d\theta = 0, \quad g \text{ in } A_0.$$
Let g_n be a sequence of elements of A (not A_0) which converge to $(1 - F)^{-1}$ in $L^2\left(\frac{1}{2\pi}d\theta\right)$. We can do this since $(1 - F)^{-1}$ is in H^2. Then
$$\int g_n f(1 - F)h\,d\theta = 0$$
for all f in A_0. Since $(1 - F)h$ is in L^2, and since g_n converges in L^2 to $(1 - F)^{-1}$, we take the limit on n and obtain
$$\int fh\,d\theta = 0, \quad \text{all } f \text{ in } A_0.$$
This proves that the absolutely continuous part of μ is orthogonal to A_0.

Theorem (F. and M. Riesz). *Let μ be a finite complex Baire measure on the unit circle such that*
$$\int e^{in\theta}d\mu(\theta) = 0, \quad n = 1, 2, 3, \ldots.$$
Then μ is absolutely continuous with respect to Lebesgue measure.

Proof. We are assuming that μ is orthogonal to A_0. If $\mu = \mu_a + \mu_s$, with μ_a absolutely continuous and μ_s singular, Corollary 2 above says that μ_a and μ_s are each orthogonal to A_0. Let's look at the singular measure μ_s. By Corollary 1 above we can find a sequence of functions f_n in A_0 which converge to 1 in L^2 of the positive singular measure $|\mu_s|$. Since μ_s is orthogonal to A_0,
$$\int d\mu_s = \lim_{n\to\infty} \int f_n d\mu_s = 0.$$
Then μ_s is also orthogonal to 1. The singular measure $e^{-i\theta}d\mu_s$ is now orthogonal to A_0; hence, it is orthogonal to 1, i.e.,
$$\int e^{-i\theta}d\mu_s(\theta) = 0.$$

So $e^{-2i\theta}d\mu_s$ is orthogonal to A_0, and, consequently, orthogonal to 1. Repeating this process, we conclude that

$$\int e^{in\theta}d\mu_s(\theta) = 0, \quad n = 0, \pm 1, \pm 2, \ldots$$

so μ_s must be the zero measure. Therefore, our original μ is absolutely continuous.

Szegö's Theorem

Now let us complete Szegö's theorem. We had reduced the problem there to proving that for an absolutely continuous positive measure $d\mu = \frac{1}{2\pi} h d\theta$ the square of the distance [in $L^2(d\mu)$] from 1 to A_0 is

$$\exp\left[\frac{1}{2\pi}\int_{-\pi}^{\pi} \log h\, d\theta\right].$$

To establish this we first prove the following.

Theorem. *Let* h *be any non-negative and Lebesgue-integrable function on the unit circle. Then*

$$\exp\left[\frac{1}{2\pi}\int_{-\pi}^{\pi} \log h(\theta)d\theta\right] = \inf_{f \in A_0} \frac{1}{2\pi}\int_{-\pi}^{\pi} h e^{\operatorname{Re} f} d\theta.$$

The left-hand side is to be interpreted as zero if log h *is not integrable.*

Proof. Since $\log h$ is bounded above by the non-negative integrable function h, we have only $\log h$ non-integrable if

$$\int \log h\, d\theta = -\infty.$$

Whether $\log h$ is integrable or not, we have

$$\exp\left[\frac{1}{2\pi}\int \log h\, d\theta\right] \leq \frac{1}{2\pi}\int h\, d\theta.$$

When h is a simple function, this is just the familiar relation between finite arithmetic and geometric means. The same inequality will apply to the function he^g if g is any real-valued integrable function. If such a g also satisfies $\int g\, d\theta = 0$, the inequality becomes

$$\exp\left[\frac{1}{2\pi}\int \log h\, d\theta\right] \leq \frac{1}{2\pi}\int h e^g d\theta.$$

In particular, this holds if $g = \operatorname{Re} f$, with f in A_0. So we have

$$\exp\left[\frac{1}{2\pi}\int \log h\, d\theta\right] \leq \inf_g \frac{1}{2\pi}\int h e^g d\theta, \quad g = \bar{g} \in L^1, \quad \int g\, d\theta = 0.$$

$$\leq \inf_{f \in A_0} \frac{1}{2\pi}\int h e^{\operatorname{Re} f} d\theta.$$

It is not difficult to see that the last two infima are equal. The functions

Re f with f in A are dense in the real continuous functions, and, consequently, dense in the real L^1 functions. Thus the functions Re f with f in A_0 are dense in the real L^1 functions g for which $\int g d\theta = 0$. This alone does not prove the equality of the two infima, since there is some difficulty when we exponentiate. The difficulty can be removed as follows. If g is a real L^1 function for which $\int g d\theta = 0$, we can find a sequence of real functions g_n in L^∞ such that $\int g_n d\theta = 0$ and g_n increases monotonely to g. By the monotone convergence theorem it is clear that the infimum is the same for the class of real L^∞ functions as it is for the class of real L^1 functions. But each real L^∞ function is the pointwise almost everywhere limit of a *bounded* sequence of the functions in Re A. Now, by the bounded convergence theorem, the infimum is the same for real L^∞ functions as it is for functions in Re A.

Now we want to reverse the inequality. Suppose first that $\log h$ is integrable. Let

$$\lambda = \frac{1}{2\pi} \int \log h d\theta$$

and put $g = \lambda - \log h$. Then g is real, in L^1, and $\int g d\theta = 0$. Also,

$$\frac{1}{2\pi} \int h e^g d\theta = \frac{1}{2\pi} \int e^\lambda d\theta = \exp\left[\frac{1}{2\pi} \int \log h d\theta\right].$$

Thus the infimum is attained at g and the inequality is actually equality.

If $\log h$ is not integrable, for any $\epsilon > 0$ the function $\log(h + \epsilon)$ is integrable, so

$$\exp\left[\frac{1}{2\pi} \int \log(h + \epsilon) d\theta\right] = \inf_{f \in A_0} \frac{1}{2\pi} \int (h + \epsilon) e^{\operatorname{Re} f} d\theta.$$

Letting ϵ tend to zero, we have the theorem.

Theorem (Szegö; Kolmogoroff-Krein). *Let μ be a finite positive Baire measure on the unit circle and let h be the derivative of μ with respect to normalized Lebesgue measure. Then*

$$\inf_{f \in A_0} \int |1 - f|^2 d\mu = \exp\left[\frac{1}{2\pi} \int_{-\pi}^{\pi} \log h(\theta) d\theta\right].$$

Proof. By the last theorem,

$$\exp\left[\frac{1}{2\pi} \int \log h d\theta\right] = \inf_{g \in A_0} \frac{1}{2\pi} \int h e^{2\operatorname{Re} g} d\theta.$$

Now $e^{2\operatorname{Re} g} = |e^g|^2$. If g is in A_0, the function e^g has the form $e^g = 1 - f$ with f in A_0, so $e^{2\operatorname{Re} g} = |1 - f|^2$. Thus

$$\exp\left[\frac{1}{2\pi} \int \log h d\theta\right] \geq \inf_{f \in A_0} \frac{1}{2\pi} \int |1 - f|^2 h d\theta.$$

We now deduce the reverse inequality by applying this one to a different function h. Let g be in A_0 and apply the last inequality with h replaced by $|1 - g|^2$. We obtain

$$\exp\left[\frac{1}{2\pi}\int \log|1-g|^2 d\theta\right] \geq \inf_{f\in A_0}\frac{1}{2\pi}\int |1-f-g+fg|^2 d\theta$$
$$\geq 1.$$

In particular, the function $\log|1-g|^2$ is Lebesgue integrable and

$$\int \log|1-g|^2 d\theta \geq 0.$$

Therefore,

$$|1-g|^2 = ke^p$$

where p is a real L^1 function such that $\int p\,d\theta = 0$ and the constant k satisfies $k \geq 1$. This tells us (returning to our original h) that

$$\frac{1}{2\pi}\int |1-g|^2 h\,d\theta = k \cdot \frac{1}{2\pi}\int he^p d\theta \geq \inf_{f\in A_0}\frac{1}{2\pi}\int he^{\operatorname{Re} f} d\theta$$
$$= \exp\left[\frac{1}{2\pi}\int \log h\,d\theta\right].$$

If we inf over g, we obtain

$$\inf_{f\in A_0}\frac{1}{2\pi}\int |1-f|^2 h\,d\theta = \exp\left[\frac{1}{2\pi}\int \log h\,d\theta\right].$$

We proved in Corollary 1 above that the infimum here is equal to

$$\inf_{f\in A_0}\int |1-f|^2 d\mu.$$

That completes the proof.

Of course, we may replace A_0 in the theorem by the family of polynomials vanishing at the origin. An immediate corollary to Szegö's theorem is that in $L^2(d\mu)$ the closed linear span of the functions $e^{in\theta}$, $n \geq 1$ contains the constant function 1 if, and only if,

$$\int \log\left(\frac{d\mu}{d\theta}\right) d\theta = -\infty.$$

This is easily seen to be equivalent to the statement that these functions span $L^2(d\mu)$.

Completion of the Discussion of H^1

Recall that the space H^p was defined as the class of analytic functions f in the open unit disc for which the functions $f_r(\theta) = f(re^{i\theta})$ are bounded in L^p-norm. For $1 < p \leq \infty$ we were rather easily able to identify H^p with the space of L^p functions on the circle such that

$$\int e^{in\theta} f(\theta) d\theta = 0, \quad n = 1, 2, 3, \ldots.$$

This used only the harmonicity of the functions in H^p. For $p = 1$, the fact that an f in H^1 was harmonic enabled us to identify f with a finite measure μ on the circle (f is the Poisson integral of μ). The fact that f was analytic told us that μ was "analytic", i.e.,

$$\int e^{in\theta} d\mu(\theta) = 0, \quad n = 1, 2, 3, \ldots.$$

Now we have the theorem of F. and M. Riesz, which tells us that $d\mu = \frac{1}{2\pi} \tilde{f} d\theta$ where $\tilde{f}$ is in L^1, and, of course, that f is the Poisson integral of $\tilde{f}$. The functions f_r converge to $\tilde{f}$ in L^1 norm, and, in fact, the non-tangential limit relation

$$\tilde{f}(\theta) = \lim_{z \to e^{i\theta}} f(z)$$

holds for almost every θ. So we may now identify H^1 with the space of L^1 functions on the circle which are "analytic", just as we did for H^p, $p > 1$.

Now we want to establish some special properties of H^p functions on the circle, chiefly the fact that if f is in H^1 then $\log |f(\theta)|$ is Lebesgue-integrable.

Theorem. *Let* f *be any function in* H^1 *such that*

$$f(0) = \frac{1}{2\pi} \int f(\theta) d\theta \neq 0.$$

Then $\log |f(\theta)|$ *is Lebesgue integrable and*

$$\frac{1}{2\pi} \int_{-\pi}^{\pi} \log |f(\theta)| d\theta \geq \log |f(0)|.$$

Proof. First suppose f is in H^2. Applying Szegö's theorem to the measure $\frac{1}{2\pi} |f|^2 d\theta$, we obtain

$$\exp \left[\frac{1}{2\pi} \int \log |f|^2 d\theta \right] = \inf_{g \in A_0} \frac{1}{2\pi} \int |1 - g|^2 |f|^2 d\theta.$$

For g in A_0,

$$\frac{1}{2\pi} \int |1 - g|^2 |f|^2 d\theta = \frac{1}{2\pi} \int |f - fg|^2 d\theta$$

$$= \frac{1}{2\pi} \int |f(0) - p|^2 d\theta$$

where p is in A_0. This last integral is not less than $|f(0)|^2$. It follows that

$$\frac{1}{2\pi} \int \log |f(\theta)| d\theta \geq \log |f(0)|.$$

If f is not in H^2, choose a sequence of functions f_n in H^2 such that $f_n(0) = f(0)$ and f_n converges to f in L^1. For example, let f_n be the nth Cesaro mean of the Fourier series for f. Then

$$\frac{1}{2\pi} \int \log |f_n| d\theta \geq \log |f(0)|.$$

Letting $n \to \infty$, we obtain the desired conclusion. In passing to the limit here, a comment may be in order. If we choose any $\epsilon > 0$, the functions $\log (|f_n| + \epsilon)$ will converge to $\log (|f| + \epsilon)$ in L^1, and we may pass to the limit with no difficulty:

$$\frac{1}{2\pi} \int \log (|f| + \epsilon) d\theta \geq \log |f(0)|.$$

Now let ϵ tend to zero, and we are done.

Corollary. *For any non-zero* f *in* H^1 *the function* $\log |f(\theta)|$ *is integrable and*

$$\frac{1}{2\pi} \int \log |f| d\theta \geq \log |f(0)|.$$

Proof. As a function on the disc we can write $f(z) = z^n g(z)$ where $g(0) \neq 0$. It is clear that g is also in H^1 and that $|g(e^{i\theta})| = |f(e^{i\theta})|$ almost everywhere. Apply the theorem to g.

Corollary. *If* f *is in* H^1, *then* f *cannot vanish on a set of positive Lebesgue measure on the circle unless* f *is identically zero.*

The inequality

$$\frac{1}{2\pi} \int \log |f| d\theta \geq \log |f(0)|$$

is simply an extension of Jensen's inequality from the case of analytic boundary values to the case of integrable boundary values. We shall discuss the analogous extension of Jensen's formula for the difference of the two quantities in the next chapter. We shall give another proof of the integrability of $\log |f|$ for f in H^1 when we discuss Dirichlet algebras. This proof will also extend to the general context of Dirichlet algebras, but it will be more elementary because it does not make use of the Szegö theorem.

One should contrast the last corollary with the situation for harmonic functions. If f is harmonic in the disc and the functions $f_r(\theta) = f(re^{i\theta})$ are bounded in L^1 norm, then f has non-tangential boundary values at almost every point on the circle. These boundary values are the derivative with respect to Lebesgue measure of the measure of which f is the Poisson integral. These boundary values may vanish almost everywhere, which happens if and only if the measure corresponding to f is singular. But when f is analytic, the non-tangential boundary values cannot vanish, even on a set of positive Lebesgue measure, unless $f = 0$.

Theorem. *Every function in* H^1 *is the product of two functions in* H^2.

Proof. It is no loss of generality to assume that we have a function f in H^1 for which $f(0) \neq 0$. Then $\log |f|$ is integrable, so for the positive measure $d\mu = \dfrac{1}{2\pi}|f|d\theta$ the constant function 1 is not in the $L^2(d\mu)$ closure of A_0, the continuous H^1 functions vanishing at the origin. By the first theorem of this chapter $|f| = k|1 - F|^{-2}$ almost everywhere, where F is the orthogonal projection of 1 onto the $L^2(d\mu)$ span of A_0, and k is a non-zero constant. By the same theorem, $(1 - F)^{-1}$ is in H^2 and $(1 - F)f$ is in L^2. It is trivial to verify that since f is in H^1 the function $(1 - F)f$ is not only in L^2 but also in H^2. The factorization of f is then

$$f = (1 - F)^{-1} \cdot (1 - F)f.$$

Theorem. *Let h be a non-negative Lebesgue-integrable function on the circle. A necessary and sufficient condition that* h *be of the form* h $= |$f$|^2$, *with* f *a non-zero function in* H^2, *is that* log h *be integrable.*

Proof. If $h = |f|^2$ with f non-zero in H^2, we know that $\log h$ is integrable. On the other hand, if $\log h$ is integrable, then $|h| = k|1 - F|^{-2}$, as in the last proof.

This theorem has another proof, which makes use of more of the special properties of the disc.

Theorem. *Let* h *be a non-negative integrable function on the circle. Then* h *is the modulus of a non-zero* H^1 *function if and only if* log h *is integrable. If* h *is non-negative and in* L^∞, *then* h *is the modulus of a non-zero* H^∞ *function if and only if* log h *is integrable.*

Proof. Both statements have been proved in one direction. Suppose $\log h$ is integrable. Let

$$f(z) = \exp\left[\frac{1}{2\pi}\int_{-\pi}^{\pi}\frac{e^{i\theta} + z}{e^{i\theta} - z}\log h(\theta)d\theta\right].$$

Then f is analytic in the disc and

$$|f| = e^u$$

where u is the Poisson integral of $\log h$. Now

$$\frac{1}{2\pi}\int_{-\pi}^{\pi}|f(re^{i\theta})|d\theta = \frac{1}{2\pi}\int_{-\pi}^{\pi}\exp[u(re^{i\theta})]d\theta.$$

Also, since $d\mu = \dfrac{1}{2\pi}P_r(\theta - t)dt$ is a positive measure of mass 1,

$$\exp[u(re^{i\theta})] = \exp\left[\frac{1}{2\pi}\int_{-\pi}^{\pi}P_r(\theta - t)\log h(t)dt\right]$$

$$\leq \frac{1}{2\pi}\int_{-\pi}^{\pi}P_r(\theta - t)h(t)dt \qquad (\exp[\textstyle\int gd\mu] \leq \int e^g d\mu)$$

and so

$$\frac{1}{2\pi}\int_{-\pi}^{\pi}|f(re^{i\theta})|d\theta \leq \frac{1}{2\pi}\int_{-\pi}^{\pi}h(t)dt.$$

Thus f is in H^1 and
$$\lim_{r \to 1} |f(re^{i\theta})| = \lim_{r \to 1} \exp[u(re^{i\theta})]$$
$$= h(\theta)$$
almost everywhere. If h is bounded, the above estimates show that f is in H^∞, i.e., it is a bounded analytic function.

Dirichlet Algebras

Let X be a compact Hausdorff space and let A be a uniformly closed complex-linear algebra of continuous complex-valued functions on X which contains the constant functions. We say that A is a **Dirichlet algebra** if the real parts of the functions in A are uniformly dense in the real continuous functions on X. The terminology is due to A. M. Gleason. Of course, we have just been working with one example of a Dirichlet algebra, the algebra of continuous functions on the unit circle whose Fourier coefficients vanish on the negative integers. Another example is the following. Let X be the torus, i.e., the product of the unit circle with itself. With each continuous function f on X, there is associated a double Fourier series
$$f(\theta, \psi) \sim \sum_{m,n} c_{mn} e^{im\theta} e^{in\psi}$$
where the Fourier coefficients c_{mn} are defined by
$$c_{mn} = \frac{1}{4\pi^2} \int_{-\pi}^{\pi} \int_{-\pi}^{\pi} f(\theta, \psi) e^{-i(m\theta + n\psi)} d\theta d\psi.$$

It is convenient to think of the Fourier coefficients as indexed by the lattice points (m, n) in the plane. If S is a set of lattice points, call S a **half-plane** if:

(i) for any two integers m, n, one and only one of the points (m, n) and $(-m, -n)$ is in S;

(ii) for (m, n) and (m_0, n_0) in S, the sum $(m + m_0, n + n_0)$ is in S.

If S is a half-plane of lattice points, let A be the set of all continuous functions on the torus whose Fourier coefficients vanish outside S. Then A is a Dirichlet algebra on the torus. There are certain obvious half-planes one might use. An interesting one is obtained by choosing an irrational number α and
$$S = \{(m, n); m + n\alpha \geq 0\}.$$

Another example (which includes the two examples above) is obtained as follows. Let X be a compact abelian group whose character group $\hat{X}$ contains a subsemigroup S, which "totally orders" $\hat{X}$:

(i) the zero (identity) of $\hat{X}$ is in S;

(ii) for any non-zero element y in $\hat{X}$, either y is in S or $-y$ is in S, not both.

Let A be the algebra of continuous functions on X whose Fourier transforms vanish outside S. Then A is a Dirichlet algebra.

In a 1958 Acta paper, Henry Helson and David Lowdenslager extended most of the theorems of this chapter to the group context described in the last example. As we said earlier, most of the elegant proofs we have given are theirs. Some of the results had been obtained for Fourier series in several variables by Bochner and others. Arens and Singer had previously done some of the complex function theory in the group context. After the appearance of the Helson-Lowdenslager paper, Bochner pointed out that their results generalize to the context of certain rings of functions. Then it was apparent that probably one natural setting for their work was Dirichlet algebras.

Let A be a Dirichlet algebra on the compact space X. If μ is a finite positive Baire measure on X, let $H^p(d\mu)$ denote the closure in $L^p(d\mu)$ of the functions in A. When X is the circle and A is our standard example, the H^p spaces are $H^p = H^p\left(\frac{1}{2\pi} d\theta\right)$. The particular relationship between A and $\frac{1}{2\pi} d\theta$ which is relevant here is that this measure is multiplicative on A:

$$\frac{1}{2\pi} \int fg \, d\theta = \frac{1}{2\pi} \int f \, d\theta \cdot \frac{1}{2\pi} \int g \, d\theta, \quad f, g \text{ in } A.$$

For the general Dirichlet algebra A we single out any non-zero positive Baire measure m on X which is *multiplicative* on A. Then we proceed to study the spaces $H^p = H^p(dm)$. We denote by A_0 the set of functions f in A such that $\int f \, dm = 0$. With the proofs given above one obtains the following results.

(1) Let μ be a positive measure on X such that 1 is not in the closed subspace of $L^2(d\mu)$ spanned by A_0. Let F be the orthogonal projection of 1 into that subspace. Then the measure $|1 - F|^2 d\mu$ is a non-zero constant multiple of dm; the function $(1 - F)^{-1}$ is in $H^2(dm)$; and if $h = d\mu/dm$ the function $(1 - F)h$ is in $L^2(dm)$.

(2) If μ is a positive measure on X and $d\mu_a = (d\mu/dm)dm$, then

$$\inf_{f \in A_0} \int |1 - f|^2 d\mu = \inf_{f \in A_0} \int |1 - f|^2 d\mu_a.$$

In particular, if μ is mutually singular with m, then 1 is in the closed subspace of $L^2(d\mu)$ spanned by A_0.

(3) If μ is a finite complex measure on X which is orthogonal to A_0, then the absolutely continuous and singular parts of μ (with respect to m) are separately orthogonal to A_0.

(4) If μ is a positive measure on X, then

$$\inf_{f \in A_0} \int |1 - f|^2 d\mu = \exp\left[\int \log\left(\frac{d\mu}{dm}\right) dm\right].$$

(5) If f is a function in $H^1(dm)$ such that $\int f dm \neq 0$, then $\log |f|$ is integrable with respect to m and

$$\int \log |f| dm \geq \log \left|\int f dm\right|.$$

(6) Every function f in $H^1(dm)$ for which $\int f dm \neq 0$ is the product of two functions in $H^2(dm)$.

(7) Let h be a non-negative function in $L^1(dm)$. Then h has the form $h = |f|^2$, where f is in $H^2(dm)$ and $\int f dm \neq 0$, if and only if $\log h$ is in $L^1(dm)$.

Some comments are in order. Statement (4) above says that the Szegö theorem is valid for Dirichlet algebras. On the other hand, the theorem of F. and M. Riesz generalizes in a different form from the circle context. We do obtain the fact that if μ is orthogonal to A_0 the absolutely continuous and singular parts of μ are orthogonal to A_0; and that any singular measure orthogonal to A_0 is also orthogonal to 1. This is where the general statement stops. It does immediately imply the classical result, because we can keep shifting the singular measure on the circle to conclude that it is zero. For the general Dirichlet algebra one can have non-zero measures orthogonal to A_0 which are mutually singular with m (see Exercise 11).

The integrability of the log of a non-zero H^1 function is false for the general Dirichlet algebra. One does have

$$\log \left|\int f dm\right| \leq \int \log |f| dm$$

so that $\log |f|$ is integrable if $\int f dm \neq 0$, but when the latter integral is zero and $f \neq 0$ it may happen that $\log |f|$ is not integrable. As we mentioned earlier, the Jensen inequality has a very short proof for Dirichlet algebras which avoids the use of the generalized Szegö theorem, and, indeed, uses only the definition of Dirichlet algebra. This inequality was first proved for Dirichlet algebras by Arens and Singer. The following short proof was communicated to me by John Wermer.

Let X, A, and m be as above. Let f be any function in A. We denote by $\hat{f}(m)$ the integral $\int f dm$. We wish to prove that

$$\int \log |f| dm \geq \log |\hat{f}(m)|.$$

Choose $\epsilon > 0$. Then $\log (|f| + \epsilon)$ is a real-valued continuous function on X. Therefore, we can find a function $g = u + iv$ in A such that

$$|u - \log (|f| + \epsilon)| < \epsilon \quad \text{on } X,$$

that is,

$$u - \epsilon < \log (|f| + \epsilon) < u + \epsilon.$$

Let $h = e^{-g}$, so that h is also in A and $|h| = e^{-u}$. On X we then have

$$|fh| = |fe^{-g}| = |f|e^{-u} < e^{\epsilon}$$

by our choice of g. Since m is multiplicative on A, we also have
$$\hat{f}(m)\hat{h}(m) = (fe^{-g})\hat{\ }(m)$$
and
$$|\hat{f}(m)|\,|\hat{h}(m)| \leq \sup_X |fe^{-g}| < e^\epsilon.$$
So
$$\log |\hat{f}(m)| + \log |\hat{h}(m)| < \epsilon$$
or
$$\log |\hat{f}(m)| - \hat{u}(m) < \epsilon.$$
Now $u < \epsilon + \log(|f| + \epsilon)$ on X; hence
$$\hat{u}(m) < \epsilon + \int \log(|f| + \epsilon)\,dm$$
and we obtain
$$\log |\hat{f}(m)| < 2\epsilon + \int \log(|f| + \epsilon)\,dm.$$
Let ϵ tend to zero, and we have Jensen's inequality for any f in A.

To prove the inequality for f in $H^1(dm)$, just approximate f in $L^1(dm)$ by functions in A.

NOTES

Szegö's theorem is in Szegö [86]. The characterization of H^1 functions and the integrability of $\log |f|$ is in F. and M. Riesz [72]. Many proofs of the F. and M. Riesz theorem on measures have been given, e.g., see Helson [42]. Some recent work on the result can be found in Bishop [10], Wermer [91], Hoffman [47], and Helson-Lowdenslager [43]. The last paper is the one from which most of the proofs in this chapter are taken. Bochner [14] pointed out the generality of their arguments. Also see Bochner's earlier paper [15]. Dirichlet algebras were defined by Gleason [34]. More about Dirichlet algebras can be found in Wermer [92, 93]. The Helson-Lowdenslager proofs extend somewhat beyond Dirichlet algebras, as Bochner observed. The critical hypotheses for the proofs given here are (i) A is an algebra of continuous complex-valued functions on the compact Hausdorff space X, containing the constant functions; (ii) m is a positive measure on X which is multiplicative on A; (iii) if μ is a positive measure on X which agrees with m on A, then $\mu = m$; (iv) the functions in A and their complex conjugates span $L^2(dm)$; (v) each real f in $L^\infty(dm)$ is the pointwise almost everywhere limit of a *bounded* sequence of functions in Re A. These hypotheses are satisfied if A is the algebra of bounded analytic functions in the unit disc, if X is the maximal ideal space of the algebra of bounded measurable functions on the unit circle, and if m is the measure on X corresponding to Lebesgue measure on the circle. Hypothesis (iii) is satisfied because of a theorem of Gleason and Whitney [35]. See Chapter 10. If one wants the results just for Dirichlet algebras, the proofs can be shortened somewhat by first proving Jensen's inequality for functions in the algebra, using the proof at the very end of this chapter. Jensen's inequality for Dirichlet algebras was first proved by Arens and Singer [5]. In fact, their proof shows that the Jensen inequality follows from hypotheses (i), (ii), and (iii) above. We should mention that hypothesis (v) above is used only in the proof of the Szegö theorem, and all of the results enumerated on pages 55–56 follow from hypotheses (i)–(iv), except possibly result (4). We shall see, in Chapter 10, other hypotheses which may replace (v) in obtaining Szegö's theorem. As a sampling of some previous generalizations of function theory in the unit disc, see Bochner [13], Mackey [57], Arens and Singer [6], and Wiener and Masani [59]. Jensen's inequality for Dirichlet algebras was first proved by Arens and Singer [5]. The integrability of log $|f|$ for certain rings of continuous functions was proved by Arens [4]. For more about analytic functions which vanish on a set of positive measure, one may consult the paper of Lusin and Privaloff [56] or Privaloff's book [70]. One result is if f is an arbitrary analytic function in the unit disc, and if there is some set of positive measure on the circle on which non-tangential limits of f exist and are zero, then f is identically zero.

EXERCISES

1. If μ is a positive measure of mass 1 and f is a real-valued function in $L^1(d\mu)$, prove that
$$\exp\left[\int f d\mu\right] \leq \int e^f d\mu$$
and equality holds if and only if $f = 0$ almost everywhere $d\mu$.

2. Let f be a non-zero function in H^1. Prove the equivalence of these two properties of f.
  (a) $\log |f(0)| = \dfrac{1}{2\pi} \int_{-\pi}^{\pi} \log |f(\theta)| d\theta$.
  (b) $f(z) = \lambda \exp\left[\dfrac{1}{2\pi} \int_{-\pi}^{\pi} \dfrac{e^{i\theta} + z}{e^{i\theta} - z} \log |f(\theta)| d\theta\right]$, where λ is a constant of modulus 1 (such an f is called an **outer** function).

3. Let μ be a finite positive Baire measure on the unit circle and suppose 1 is *not* in the $L^2(d\mu)$ closure of A_0, the analytic functions with continuous boundary values which vanish at the origin. Let F be the orthogonal projection of 1 into the $L^2(d\mu)$ closure of A_0 and let $G = (1 - F)^{-1}$. Prove the following.
  (a) G is in H^2; G is an outer function (Exercise 2); $G(0) = 1$.
  (b) The absolutely continuous part of μ is
$$d\mu_a = \frac{k}{2\pi} |G|^2 d\theta,$$
where $k = \inf\limits_{f \in A_0} \int |1 - f|^2 d\mu$.
  (c) If E is a Baire set such that μ_a lives on E and μ_s (the singular part of μ) lives on the complement of E, then the characteristic function of E is in $H^2(d\mu)$, the closure of A in $L^2(d\mu)$.
  (d) With reasonable conventions
$$H^2(d\mu) = H^2(d\mu_a) \oplus L^2(d\mu_s).$$

4. Let μ be a finite positive Baire measure on the unit circle. Prove the equivalence of the following.
  (a) 1 is *not* in the $L^2(d\mu)$ closure of A_0.
  (b) $H^2(d\mu) \neq L^2(d\mu)$.

(c) If we norm A with the $L^2(d\mu)$ norm, evaluation at the origin is a bounded linear functional on A.

(d) For each z in the open disc "evaluation at z" is a bounded linear functional on A, using the $L^2(d\mu)$ norm on A.

(e) For each function f in $H^2(d\mu)$ there is an analytic function g in the unit disc (which is the quotient of two ordinary H^2 functions) such that the non-tangential limits of g agree with f almost everywhere with respect to Lebesgue measure.

5. For each α, $|\alpha| < 1$, evaluation at α is a bounded linear functional on the Hilbert space H^2. This functional is, therefore, "inner product with" some function in H^2. Which function?

6. Use Szegö's theorem to prove the following. If f is a function in L^2 of the circle, then the functions $e^{in\theta}f(\theta)$, $n \geq 0$, span L^2 of the circle if and only if
  (a) f does not vanish on a set of positive Lebesgue measure;
  (b) $\log |f|$ is *not* Lebesgue integrable.

7. Let μ_1 and μ_2 be positive measures on the circle. In each of the L^2 spaces we complete the polynomials in z, arriving at the Hilbert spaces $H^2(d\mu_1)$, $H^2(d\mu_2)$. On each of these spaces we consider the linear operator "multiplication by z." Prove that these two multiplication operators are unitarily equivalent if and only if
  (a) μ_1 and μ_2 are mutually absolutely continuous.
  (b) the functions $\log(d\mu_j/d\theta)$, $j = 1, 2$, are either both Lebesgue integrable or both not Lebesgue integrable.

8. Let f be in L^p of the unit circle C, $1 < p < \infty$, and define g in the open unit disc by
$$g(z) = \frac{1}{2\pi i} \int_C \frac{f(\lambda)}{\lambda - z} d\lambda.$$
Is g in H^p?

9. Prove that for any f analytic in $|z| < 1$ the L^p norms
$$\frac{1}{2\pi} \int_{-\pi}^{\pi} |f(re^{i\theta})|^p d\theta$$
are increasing in r.

10. If τ is a conformal map of the disc onto the disc, does $f(z) \to f(\tau(z))$ map H^2 into itself?

11. Consider the torus $T = \{(e^{i\theta}, e^{i\psi})\}$ and Lebesgue measure $dm = \frac{1}{4\pi^2} d\theta d\psi$ thereon.
  (a) Prove that the functions $e^{ik\theta}$, $e^{in\psi}$ are an orthonormal family in $L^2(dm)$.
  (b) Find a Fejer kernel, giving the Cesaro means of the Fourier series for a function in $L^1(dm)$. Prove the analogue of Fejer's theorem, the completeness of the functions in part (a), etc.
  (c) Choose an irrational number α and let A consist of all continuous functions of the torus whose Fourier coefficients a_{kn} vanish for $k + n\alpha < 0$. Prove that A is a Dirichlet algebra on torus.

(d) Show that the F. and M. Riesz theorem is false for A. That is, let A_0 be the set of functions in A which vanish at the origin:
$$\int_T f dm = 0.$$
Then show that the measure μ defined by
$$\int_T f d\mu = \int_{-\infty}^{\infty} f(e^{-it}, e^{-i\alpha t})(1-it)^{-2} dt, \qquad f \in C(T)$$
is singular, non-zero, and orthogonal to A_0.

CHAPTER 5

FACTORIZATION FOR H^p FUNCTIONS

Inner and Outer Functions

Suppose f is a non-zero function of the class H^1 on the unit disc. Then f has non-tangential limits at almost every point of the unit circle:

$$f(e^{i\theta}) = \lim_{z \to e^{i\theta}} f(z)$$

and

$$f(re^{i\theta}) = \frac{1}{2\pi} \int_{-\pi}^{\pi} f(e^{it}) P_r(\theta - t) dt.$$

Also, $\log |f(e^{it})|$ is Lebesgue integrable. Let

$$F(z) = \exp\left[\frac{1}{2\pi} \int_{-\pi}^{\pi} \frac{e^{i\theta} + z}{e^{i\theta} - z} \log |f(e^{i\theta})| d\theta\right].$$

Then F is an analytic function in the unit disc. Also, F is in H^1 because

$$\frac{1}{2\pi} \int_{-\pi}^{\pi} |F(re^{i\theta})| d\theta \leq \frac{1}{2\pi} \int_{-\pi}^{\pi} |f(e^{i\theta})| d\theta.$$

This results from the fact that $|F| = e^u$, where u is the Poisson integral of $\log |f|$. Clearly, $|F| = |f|$ almost everywhere on the unit circle. Of course, F has no zeros in the open disc, and

$$\log |F(re^{i\theta})| = \frac{1}{2\pi} \int_{-\pi}^{\pi} \log |f(e^{it})| P_r(\theta - t) dt.$$

Thus, $|F(z)| \geq |f(z)|$ for each z in the open disc. For

$$\log |f(re^{i\theta})| \leq \frac{1}{2\pi} \int_{-\pi}^{\pi} \log |f(e^{it})| P_r(\theta - t) dt = \log |F(re^{i\theta})|.$$

This inequality is just an application of Jensen's inequality as we proved it in the last chapter, but using the measure

$$dm = \frac{1}{2\pi} P_r(\theta - t) dt$$

instead of $\frac{1}{2\pi} dt$. It is just the statement that $\log |f|$ is subharmonic; i.e., lies everywhere below the harmonic function $\log |F|$, which has the same boundary values as $\log |f|$.

Let us look at the function

$$g(z) = \frac{f(z)}{F(z)}.$$

Now g is a bounded analytic function in the open disc; indeed,

$$|g(z)| = \left|\frac{f(z)}{F(z)}\right| \leq 1$$

in the disc. Also, $|g(e^{i\theta})| = 1$ almost everywhere on the unit circle. Thus we have f as the product $f = gF$ of a bounded analytic function of modulus one on the boundary and an H^1 function F of a rather special type. It will be convenient for us to make the following definitions.

An **inner function** is an analytic function g in the unit disc such that $|g(z)| \leq 1$ and $|g(e^{i\theta})| = 1$ almost everywhere on the unit circle. An **outer function** is an analytic function F in the unit disc of the form

$$F(z) = \lambda \exp\left[\frac{1}{2\pi}\int_{-\pi}^{\pi} \frac{e^{i\theta} + z}{e^{i\theta} - z} k(\theta) d\theta\right]$$

where k is a real-valued integrable function on the circle and λ is a complex number of modulus 1. It is easy to see that such an outer function F is in H^1 if, and only if, e^k is also integrable; when F is an outer function in H^1 we have necessarily

$$k(\theta) = \log |F(e^{i\theta})| \text{ a.e.}$$

The following characterizations of outer functions are useful.

Theorem. *Let F be a non-zero function in* H^1. *The following are equivalent.*

(i) *F is an outer function.*

(ii) *If* f *is any function in* H^1 *such that* $|f| = |F|$ *almost everywhere on the unit circle, then* $|F(z)| \geq |f(z)|$ *at each point* z *in the open unit disc.*

(iii) $\log |F(0)| = \frac{1}{2\pi}\int_{-\pi}^{\pi} \log |F(e^{i\theta})| d\theta.$

Proof. We gave above the proof that (i) implies (ii). On the other hand, if (ii) holds, let G be the outer function

$$G(z) = \exp\left[\frac{1}{2\pi}\int_{-\pi}^{\pi} \frac{e^{i\theta} + z}{e^{i\theta} - z} \log |F(e^{i\theta})| d\theta\right].$$

Then $|F(z)| \leq |G(z)| \leq |F(z)|$ on the disc. Thus, F/G is analytic and everywhere of absolute value 1. So $F = \lambda G$ where $|\lambda| = 1$, proving that F is an outer function. Obviously, (iii) holds for any outer function in H^1. Suppose that (iii) holds. Define G as above, and we see that F/G is bounded by 1 in the disc and has absolute value 1 at $z = 0$. Thus $F/G = \lambda$, a constant of modulus 1.

Theorem. *Let* f *be a non-zero function in* H^1. *Then* f *can be written in the form* $f = gF$ *where* g *is an inner function and* F *is an outer function. This factorization is unique up to a constant of modulus 1 and (of course) the outer function* F *is in* H^1.

Proof. As we observed above, if

$$F(z) = \exp\left[\frac{1}{2\pi}\int_{-\pi}^{\pi}\frac{e^{i\theta}+z}{e^{i\theta}-z}\log|f(e^{i\theta})|d\theta\right]$$

then F is an outer function in H^1, and $f/F = g$ is an inner function. If we also have $f = g_1 F_1$ with g_1 inner and F_1 outer, then $|F| = |F_1|$ on the boundary. Clearly, then, $F = \lambda F_1$ for some number λ of modulus 1. So $\lambda g F_1 = g_1 F_1$ and $g_1 = \lambda g$.

Blaschke Products and Singular Functions

We are now going to factor each inner function into a product of two more specialized inner functions. The first of these will be a Blaschke product, to take care of the zeros of the given inner function. The second factor will be determined by a singular measure on the unit circle.

Theorem. *Let* f *be a bounded analytic function in the unit disc and suppose* $f(0) \neq 0$. *If* $\{\alpha_n\}$ *is the sequence of zeros of* f *in the open disc, each repeated as often as the multiplicity of the zero of* f, *then the product* $\prod_n |\alpha_n|$ *is convergent, that is,*

$$\Sigma\,(1 - |\alpha_n|) < \infty.$$

Proof. Suppose $|f| \leq 1$, for convenience. Of course, f may have only a finite number of zeros, and then there is no question of convergence of their product. Otherwise, f has a countable number of zeros: $\alpha_1, \alpha_2, \alpha_3, \ldots$ and we are going to prove that the infinite product

$$\prod_{n=1}^{\infty} |\alpha_n|$$

converges.

Let $B_n(z)$ be the finite product:

$$B_n(z) = \prod_{k=1}^{n}\frac{z-\alpha_k}{1-\bar{\alpha}_k z}.$$

Now $B_n(z)$ is a rational function, analytic in the closed unit disc, and $|B_n(e^{i\theta})| = 1$, since each of the functions

$$\frac{z-\alpha_k}{1-\bar{\alpha}_k z}$$

is of modulus 1 on the unit circle. Furthermore, f/B_n is a bounded analytic function in the disc. Since

$$\frac{|f(e^{i\theta})|}{|B_n(e^{i\theta})|} = |f(e^{i\theta})| \leq 1, \quad \text{a.e.}$$

we have $|f(z)| \leq |B_n(z)|$ on the disc. In particular,

$$0 < |f(0)| \leq |B_n(0)| = \prod_{k=1}^{n} |\alpha_k|.$$

Since $|\alpha_k| < 1$ for each k, and since each of the partial products $\prod_{k=1}^{n} |\alpha_k|$ is not less than $|f(0)|$, the infinite product converges, completing the proof.

What we have in mind here is to form an infinite product from the zeros of f, thus obtaining a function by which we can divide f to arrive at a function with no zeros. Thus one would hope that the infinite product

$$\prod_{n=1}^{\infty} \frac{z - \alpha_n}{1 - \bar{\alpha}_n z}$$

should converge; however, it need not. Fortunately, if we simply rotate the nth term in the product by $-\bar{\alpha}_n/|\alpha_n|$, the new infinite product does converge.

Lemma. *Let $\{\alpha_n\}$ be a sequence of non-zero complex numbers in the open unit disc. A necessary and sufficient condition that the infinite product*

$$\prod_{n=1}^{\infty} \frac{\bar{\alpha}_n}{|\alpha_n|} \frac{\alpha_n - z}{1 - \bar{\alpha}_n z}$$

should converge uniformly on compact subsets of the disc is that the product $\prod |\alpha_n|$ should converge, i.e., that

$$\sum_{n=1}^{\infty} (1 - |\alpha_n|) < \infty.$$

When this condition is satisfied, the product defines an inner function whose zeros are exactly $\alpha_1, \alpha_2, \ldots$.

Proof. Let us first prove the last statement. Form the partial products

$$B_n(z) = \prod_{k=1}^{n} \frac{\bar{\alpha}_k}{|\alpha_k|} \frac{\alpha_k - z}{1 - \bar{\alpha}_k z}.$$

Each B_n is analytic in the closed unit disc and has modulus 1 on the unit circle. If the sequence $\{B_n\}$ converges uniformly on compact subsets of $|z| < 1$ to a function B, it is clear that B is bounded by 1 and is analytic in the interior of the unit disc. Of course, uniform convergence of the infinite product means more than uniform convergence of the B_n. It means that on each compact subset of the open disc at most a finite number of the factors in the product have a zero, and that when these factors are removed the partial products of the remaining infinite product converge uniformly to a function without zeros on the compact set. Uniform convergence of the infinite product on compact sets certainly yields a bounded analytic function whose zeros are $\alpha_1, \alpha_2, \ldots$. In particular, this convergence implies

$$\sum_{n=1}^{\infty} (1 - |\alpha_n|) < \infty.$$

Now suppose the α_n satisfy this last condition. Certainly, then, no more than a finite number of factors in the infinite product can have a zero on any given compact set. We wish to establish uniform convergence of the product on each closed disc $|z| < r < 1$. Let

$$f_n(z) = \left(\frac{\bar{\alpha}_n}{|\alpha_n|}\right) \frac{\alpha_n - z}{1 - \bar{\alpha}_n z}.$$

Then

$$1 - f_n(z) = \frac{1}{|\alpha_n|}\left[1 - \frac{|\alpha_n|^2 - \bar{\alpha}_n z}{1 - \bar{\alpha}_n z}\right] + 1 - \frac{1}{|\alpha_n|}$$

$$= \frac{1 - |\alpha_n|}{|\alpha_n|}\left[\frac{1 + |\alpha_n|}{1 - \bar{\alpha}_n z} - 1\right].$$

If $|z| \leq r$, then

$$|1 - f_n(z)| \leq \frac{1 - |\alpha_n|}{|\alpha_n|}\left[\frac{2}{1 - r} + 1\right].$$

Since $\Sigma (1 - |\alpha_n|) < \infty$, we see that $\Sigma |1 - f_n(z)|$ is uniformly summable on $|z| \leq r$, and hence that the product $\Pi f_n(z)$ is uniformly and absolutely convergent on that disc.

The convergence of the infinite product, if it is assumed that $\Sigma (1 - |\alpha_n|) < \infty$, has another proof which is interesting and should be mentioned. This proof also shows easily that the product is an inner function. Let B_n be the nth partial product as above. It is then easy to see that $\{B_n\}$ converges in H^2 on the circle. For

$$\frac{1}{2\pi}\int_{-\pi}^{\pi}|B_m - B_n|^2 d\theta = \frac{1}{2\pi}\int_{-\pi}^{\pi}[|B_m|^2 + |B_n|^2 - 2\operatorname{Re} B_n \bar{B}_m]d\theta.$$

Since each B_k has modulus 1 on the circle,

$$|B_m|^2 = |B_n|^2 = 1 \quad \text{and} \quad \bar{B}_m = \frac{1}{B_m}.$$

Thus

$$\frac{1}{2\pi}\int_{-\pi}^{\pi}|B_m - B_n|^2 d\theta = 2\left[1 - \operatorname{Re}\frac{1}{2\pi}\int_{-\pi}^{\pi}\frac{B_n}{B_m}d\theta\right].$$

If $n > m$, then B_n/B_m is analytic and

$$\frac{1}{2\pi}\int_{-\pi}^{\pi}\frac{B_n}{B_m}d\theta = \left(\frac{B_n}{B_m}\right)(0) = \prod_{k=m+1}^{n}|\alpha_k|.$$

Thus
$$\frac{1}{2\pi}\int_{-\pi}^{\pi}|B_m - B_n|^2 d\theta = 2\left(1 - \prod_{k=m+1}^{n}|\alpha_k|\right).$$

Since the infinite product $\prod |\alpha_k|$ converges, we see that $B_n \to B$ in H^2. This L^2 convergence on the boundary easily yields uniform convergence of B_n to B on compact subsets of the disc. A subsequence of the B_n converges pointwise almost everywhere to B on the circle, from which it is clear that B has modulus 1; i.e., B is an inner function.

A **Blaschke product** is an analytic function B of the form

$$B(z) = z^p \prod_{n=1}^{\infty}\left[\frac{\bar{\alpha}_n}{|\alpha_n|} \cdot \frac{\alpha_n - z}{1 - \bar{\alpha}_n z}\right]^{p_n}$$

where

(i) $p, p_1, p_2, \ldots$ are non-negative integers;
(ii) the α_n are distinct non-zero numbers in the open unit disc;
(iii) the product $\prod_n |\alpha_n|^{p_n}$ is convergent.

We have just seen that such a product converges uniformly on compact sets and that the only zeros of B are a zero of order p at the origin and a zero of order p_n at α_n. Of course, if $p = 0$ or $p_n = 0$, the corresponding term in the product may be deleted so that one retains only the factors which give rise to zeros. The only reason for allowing the orders to be **0** is to have one unified definition of Blaschke product.

Theorem. *Let f be a non-zero bounded analytic function in the unit disc. Then f is uniquely expressible in the form f = Bg where B is a Blaschke product and g is a (necessarily bounded) analytic function without zeros.*

Proof. Since $f \neq 0$, we can write $f(z) = z^p h(z)$ where $h(0) \neq 0$. Let B be the product of z^p and the Blaschke product formed from the zeros of h. Then $g = f/B$ is analytic and bounded in the disc. The factorization $f = Bg$ is unique, since a Blaschke product is uniquely determined by its zeros.

Suppose we apply this last theorem when f is an inner function. Then we shall have $f = Bg$, where B is a Blaschke product and g is an inner function without zeros.

Theorem. *Let g be an inner function without zeros, and suppose that g(0) is positive. Then there is a unique singular positive measure μ on the unit circle such that*

$$g(z) = \exp\left[-\int \frac{e^{i\theta} + z}{e^{i\theta} - z}\, d\mu(\theta)\right].$$

Proof. Since g is analytic in the disc and has no zeros, $g = e^{-h}$, where h

is an analytic function in the disc. Since g is bounded by 1, the real part of h must be non-negative on the disc. Let $h = u + iv$ so that $u \geq 0$. The non-negative harmonic function u is uniquely expressible in the form

$$u(re^{i\theta}) = \int P_r(\theta - t) d\mu(t)$$

where μ is a positive measure on the circle. Since $g(0) > 0$, we have $v(0) = 0$ (or, at least, we may assume so by subtracting $2k\pi i$ from h). Thus,

$$h(z) = \int \frac{e^{i\theta} + z}{e^{i\theta} - z} d\mu(\theta).$$

Now $|g| = 1$ almost everywhere on the circle. Since $|g| = e^{-u}$, this just means that the non-tangential limits of u must vanish almost everywhere on the circle. But these non-tangential limits are equal to $\frac{1}{2\pi} \frac{d\mu}{d\theta}$. So μ is singular, and that completes the proof.

One important part of the above proof is the fact that if h is an analytic function with values in the right half-plane and $h(0) > 0$, then

$$h(z) = \int \frac{e^{i\theta} + z}{e^{i\theta} - z} d\mu(\theta)$$

for some positive measure μ on the circle. This is usually known as Herglotz's theorem. It is equivalent to the theorem that a non-negative harmonic function in the disc is the Poisson integral of a positive measure, this also being known as Herglotz's theorem.

Let us call an inner function without zeros which is positive at the origin a **singular function.**

The Factorization Theorem

Theorem. *Let* f $\neq 0$ *be an* H¹ *function in the unit disc. Then* f *is uniquely expressible in the form* f = BSF, *where* B *is a Blaschke product,* S *is a singular function, and* F *is an outer function (in* H¹).

Proof. We know that $f = gF$, where g is an inner function and F is an outer function, and that this factorization is unique up to a constant multiple of modulus 1. If B is the Blaschke product formed from the zeros of g (i.e., the zeros of f) then $g = BS$, where S is an inner function without zeros. By multiplying g by a constant of modulus 1, we can arrange that $S(0) > 0$, i.e., that S is a singular function. We then absorb that constant into the outer function F, and we are done.

Let us make a final description of the factorization $f = BSF$. Let p be the order of the zero of f at the origin, and let $\alpha_1, \alpha_2, \ldots$ be the remaining zeros of f, the multiplicity of α_n being p_n. Then

$$B(z) = z^p \prod_{n=1}^{\infty} \left[\frac{\bar{\alpha}_n}{|\alpha_n|} \cdot \frac{\alpha_n - z}{1 - \bar{\alpha}_n z} \right]^{p_n}$$

$$F(z) = \exp\left[\frac{1}{2\pi} \int_{-\pi}^{\pi} \frac{e^{i\theta} + z}{e^{i\theta} - z} (\log|f(e^{i\theta})| + ia) d\theta \right]$$

where $a = \arg(f/B)(0)$; and then

$$S(z) = \frac{f(z)}{B(z) F(z)} = \exp\left[-\int \frac{e^{i\theta} + z}{e^{i\theta} - z} d\mu(\theta) \right]$$

for some positive singular measure μ. From this one can deduce, among other things, a generalized Jensen formula. If $f(0) \neq 0$, then

$$\frac{1}{2\pi} \int_{-\pi}^{\pi} \log|f(e^{i\theta})| d\theta = \log|f(0)| + \sum_n p_n \log|\alpha_n|^{-1} + \int d\mu.$$

If f is analytic in the closed disc, then the "singular part" of f is zero, B is a finite Blaschke product, and one has the usual Jensen formula.

Of course, we shall call F the **outer part** of f and $B \cdot S$ the **inner part** of f. We know that F is in H^1 and that B and S are bounded. One should also note that Blaschke products and singular functions are analytic in a much larger region than the open unit disc. For emphasis, we might state these as theorems.

Theorem. *The Blaschke product whose zeros are*

$$\alpha_1, \alpha_2, \alpha_3, \ldots, \quad 0 < |\alpha_n| < 1$$

converges at all points z in the complex plane except those in the compact set K *consisting of*

(i) *the points* $z = 1/\bar{\alpha}_n$;
(ii) *the points* z *on the unit circle which are accumulation points of the sequence* $\{\alpha_n\}$.

The convergence is uniform on any closed set in the plane which is disjoint from K, *and the product* B(z) *is thus analytic off* K.

Proof. On any closed set disjoint from K, the numbers $|1 - \bar{\alpha}_n z|$ are uniformly bounded away from zero. The same estimates used for convergence in the disc then apply.

We should remark that $B(z)$ has a pole at each $1/\bar{\alpha}_n$, and has an essential singularity at the accumulation points of the α_n. In particular, B cannot be extended continuously from the interior of the disc to any such accumulation point, for the extended value of B would have to be zero, while the non-tangential limits of B are of modulus 1 almost everywhere.

Theorem. *If* S *is the singular function determined by the positive singular measure* μ, *then* S *is analytic everywhere in the complex plane except at those points of the unit circle which are in the closed support of the measure* μ.

The function S (or even |S|) is not continuously extendable from the interior of the disc to any point in the closed support of μ.

Proof. The closed support of μ is the complement of the union of all open sets on the circle which have μ-measure zero. Let K be this closed support set. If z is not in K, then the function

$$h(z) = \int \frac{e^{i\theta} + z}{e^{i\theta} - z} d\mu(\theta)$$

is analytic at z with derivative

$$h'(z) = \int \frac{2e^{i\theta}}{(e^{i\theta} - z)^2} d\mu(\theta).$$

$S = e^{-h}$ is analytic off K. This has nothing to do with the fact that μ is singular.

Suppose we have a point $e^{i\theta_0}$ on the circle such that $|S|$ extends continuously to this point. Certainly, then, $|S(e^{i\theta_0})| = 1$, so Re h must be bounded near this point. That is, there is a positive δ such that Re $h(re^{i\theta})$ is bounded for $|\theta - \theta_0| \leq \delta$. Let $u =$ Re h so that

$$u(r, \theta) = \int P_r(\theta - t) d\mu(t).$$

Let g_r be the restriction of u_r to the interval $|\theta - \theta_0| \leq \delta$, and then $\{g_r\}$ is a bounded family of continuous functions on that interval. Thus, there exists a bounded measurable function g which is a weak-star cluster point of this family. Let F be any continuous function on the circle which vanishes off the interval $|\theta - \theta_0| \leq \delta$. Then

$$\lim_{r \to 1} \frac{1}{2\pi} \int_{-\pi}^{\pi} F(\theta) u(r, \theta) d\theta = \int F d\mu$$

and the integrals

$$\frac{1}{2\pi} \int_{\theta_0 - \delta}^{\theta_0 + \delta} F(\theta) u(r, \theta) d\theta$$

cluster at

$$\frac{1}{2\pi} \int_{\theta_0 - \delta}^{\theta_0 + \delta} F(\theta) g(\theta) d\theta.$$

Since F vanishes for $|\theta - \theta_0| > \delta$, we see that the last integral must be equal to $\int F d\mu$. Since this holds for all such F, we see that the restriction of μ to $|\theta - \theta_0| < \delta$ is absolutely continuous with respect to Lebesgue measure on that interval and $\frac{d\mu}{d\theta} = \frac{1}{2\pi} g$. Since μ is singular, $g = 0$ and $e^{i\theta_0}$ is not in the closed support of μ. That completes the proof.

Theorem. *Let* f *be a function in* H¹. *Then* f *is in* H^p, $1 \leq p \leq \infty$, *if and only if the outer part of* f *is in* H^p. *If* f *is continuous on the unit circle, so is the outer part of* f.

Proof. The statement about H^p functions is obvious. Suppose f is continuous on the circle. Let K be the closed set of points on the circle at which $f(e^{i\theta}) = 0$. Let $f = BSF$ be the canonical factorization of f. Certainly, every point of accumulation of the zeros of the Blaschke product B is contained in K, so B is analytic (and non-zero) at every point of the circle which is not in K. The outer function

$$F(z) = \exp\left[\frac{1}{2\pi}\int_{-\pi}^{\pi}\frac{e^{i\theta}+z}{e^{i\theta}-z}(\log|f(e^{i\theta})| + i a)d\theta\right]$$

is continuous (and zero) on K because $|F| = |f|$. Since $|F|$ and B are continuous off K, so is $|S|$. By the last theorem, then, the measure μ which determines S must have its closed support contained in K. Consequently, S is analytic off K. It follows that $F = f/BS$ is continuous off K, and hence that F is continuous on all of the circle.

Absolute Convergence of Taylor Series

This is a short section which contains two interesting theorems. One is the theorem of Hardy and Littlewood which states that if a function in H^1 is of bounded variation on the unit circle then the Taylor series for the function is absolutely convergent. The other is Hardy's theorem on the growth of the Fourier coefficients of an H^1 function. We treat the latter theorem first.

The Riemann-Lebesgue lemma states that the Fourier coefficients of an integrable function tend to zero. For H^1 functions, one can say much more.

Theorem (Hardy). *Let f be a function in* H^1 *with power series*

$$\sum_{n=0}^{\infty} a_n z^n.$$

Then

$$\sum_{n=1}^{\infty} \frac{1}{n}|a_n| \leq \pi \|f\|_1.$$

Proof. First suppose that $a_n \geq 0$, $n = 0, 1, 2, \ldots$. Then

$$\operatorname{Im} f(re^{i\theta}) = \sum_{n=1}^{\infty} a_n r^n \sin n\theta.$$

Since

$$\frac{1}{2\pi}\int_0^{2\pi}(\pi-\theta)\sin n\theta\, d\theta = \frac{1}{n}$$

we obtain

$$\sum_{n=1}^{\infty}\frac{1}{n}a_n r^n = \frac{1}{2\pi}\int_0^{2\pi}(\pi-\theta)\operatorname{Im} f(re^{i\theta})d\theta$$

$$\leq \tfrac{1}{2}\int_0^{2\pi}|f(re^{i\theta})|d\theta \leq \pi\|f\|_1.$$

Let r tend to 1 and we have what we want, assuming that $a_n \geqq 0$. For the general f write $f = gh$ where g and h are the H^2 functions defined by

$$g = B\left(\frac{f}{B}\right)^{1/2}, \qquad h = \left(\frac{f}{B}\right)^{1/2},$$

and B is the Blaschke product of the zeros of f. If

$$g(z) = \sum_{n=0}^{\infty} b_n z^n$$

$$h(z) = \sum_{n=0}^{\infty} c_n z^n$$

then by the Riesz-Fischer theorem the functions

$$G(z) = \Sigma |b_n| z^n$$

$$H(z) = \Sigma |c_n| z^n$$

are also in H^2; in fact,

$$||G||_2 = ||g||_2 \quad \text{and} \quad ||H||_2 = ||h||_2.$$

Let $F = GH$. Certainly, F is in H^1, and

$$F(z) = \sum_{n=0}^{\infty} \tilde{a}_n z^n$$

where $\tilde{a}_n \geqq 0$. It is also apparent that $|a_n| \leqq \tilde{a}_n$. By the first part of our proof

$$\sum_{n=1}^{\infty} \frac{1}{n} |a_n| \leqq \sum_{n=1}^{\infty} \frac{1}{n} \tilde{a}_n \leqq \pi ||F||_1.$$

But

$$||F||_1 \leqq ||G||_2 \, ||H||_2 = ||g||_2 \, ||h||_2 = ||f||_1$$

and we are done.

Theorem. *Let* f *be a function on the unit circle which is both of bounded variation and in* H^1. *Then*
  (i) f *is an absolutely continuous function;*
  (ii) *the Fourier series for* f *is absolutely convergent.*

Proof. Since f is of bounded variation, the Fourier coefficients of f are

$$a_n = \frac{1}{2\pi} \int_{-\pi}^{\pi} f(\theta) e^{-in\theta} d\theta = \frac{i}{n} \cdot \frac{1}{2\pi} \int_{-\pi}^{\pi} e^{-in\theta} df(\theta) \quad (n \neq 0).$$

In particular, df is analytic. By the theorem of F. and M. Riesz, df is absolutely continuous, i.e., $df = g d\theta$, where g is in H^1. Thus $a_n = \frac{i}{n} b_n$, where b_n is the nth Fourier coefficient of g, $n = 1, 2, 3, \ldots$. By the last theorem

$$\sum_{n=1}^{\infty} |a_n| = \sum_{n=1}^{\infty} \frac{1}{n} |b_n| < \infty.$$

That completes the argument.

Remarks on the Classical Approach

Let us make a few remarks on the more standard approach to the theorems of this and the last chapter. Having defined the classes H^p in the unit disc, $p > 0$, one can proceed directly to dispose of the case $p > 1$. We did this by using the weak-star compactness of the unit ball in L^p. Needless to say, the case $p = 2$ is much more easily disposed of; for if

$$f(z) = \sum_{n=0}^{\infty} a_n z^n$$

then

$$\frac{1}{2\pi} \int_{-\pi}^{\pi} |f(re^{i\theta})|^2 d\theta = \sum_{n=0}^{\infty} |a_n|^2 r^{2n}.$$

So if f is in H^2, we let $r \to 1$ and conclude that $\{a_n\}$ is square-summable. By the Riesz-Fischer theorem, the a_n are the Fourier coefficients of an L^2 function on the circle, and it is apparent that f is the Poisson integral of that function. One can then prove Fatou's theorem as we did, obtaining non-tangential convergence almost everywhere of f to its boundary values. If our function f happens to be bounded, the non-tangential convergence makes it apparent that the boundary values of f define a bounded measurable function. This settles at least H^∞ and H^2 as far as the boundary value theorems are concerned.

For any $p > 0$ and any analytic f, it is relatively easy to show that the L^p norms of the functions $f_r(\theta) = f(re^{i\theta})$ are increasing as $r \to 1$. From this one can see, for an f in H^p, that the infinite product of the moduli of the zeros of f converges. Thus, any f in H^p is of the form $f = Bg$, where B is a Blaschke product and g is a function in H^p without zeros. Since g has no zeros, $g^{p/2}$ is analytic and is in H^2. Thus, $g^{p/2}$ has non-tangential limits almost everywhere. Consequently, such limits also exist for

$$f = B(g^{p/2})^{2/p}.$$

In particular, if f is in H^1, we have

$$f = (B\sqrt{g})\sqrt{g}$$

i.e., f is the product of two H^2 functions. We thereby obtain the theorem of F. and M. Riesz that every function in H^1 is the Poisson integral of an L^1 function on the circle, that L^1 function being the product of the boundary values of $B\sqrt{g}$ and those of $\sqrt{g}$.

The boundary values of a function in H^p define an L^p function on the circle, and for any $p \geq 1$ the Poisson integral of this function is f. Of course, one would not expect such an integral representation for $p < 1$. The integrability of $\log |f(e^{i\theta})|$ for f in H^p is fairly easy to obtain. If one defines for $h \geq 0$

$$\log^+ h = \max [\log h, 0]$$
$$\log^- h = \min [\log h, 0]$$

one has the following. Since $\log x \leq x^\alpha$ for large x and a given positive α, it is easy to see that for any f in H^p the integrals

$$\frac{1}{2\pi} \int_{-\pi}^{\pi} \log^+ |f(re^{i\theta})| d\theta$$

are bounded as $r \to 1$. By the classical Jensen inequality

$$\frac{1}{2\pi} \int_{-\pi}^{\pi} \log^+ |f(re^{i\theta})| d\theta + \frac{1}{2\pi} \int_{-\pi}^{\pi} \log^- |f(re^{i\theta})| d\theta \geq \log |f(0)|.$$

Assuming $f(0) \neq 0$, one sees that the integrals

$$\frac{1}{2\pi} \int_{-\pi}^{\pi} \log |f(re^{i\theta})| d\theta$$

are bounded as $r \to 1$. From this the integrability of $\log |f(e^{i\theta})|$ is rather easy to obtain.

The factorization theory for H^1 may now proceed as we did it above.

Functions of Bounded Characteristic

The factorization we have given for H^p functions generalizes to the class of functions of **bounded characteristic**, i.e., meromorphic functions which are the quotient of two bounded analytic functions. If $f = g/h$, where g and h are bounded analytic functions with h not identically zero, it is apparent that f has finite non-tangential limits at almost every point of the circle. Suppose we write g and h as products of Blaschke products, singular functions, and outer functions, say $g = B_g S_g O_g$, $h = B_h S_h O_h$. The quotient O_g/O_h has the form

$$\lambda \exp \left\{ \frac{1}{2\pi} \int_{-\pi}^{\pi} \frac{e^{i\theta} + z}{e^{i\theta} - z} k(\theta) d\theta \right\}$$

where λ is a scalar of modulus 1 and

$$k = \log |g| - \log |h| = \log |f|.$$

It is, therefore, an outer function (though not necessarily in H^1).

The quotient S_g/S_h has the form

$$\exp \left\{ \int \frac{e^{i\theta} + z}{e^{i\theta} - z} d\mu(\theta) \right\}$$

where μ is a real singular measure; i.e., $\mu = \mu_h - \mu_g$ where these are the positive singular measures going with h and g.

We conclude that every function of bounded characteristic has the form

$$f(z) = \lambda \frac{B_1(z)}{B_2(z)} \exp\left\{\int \frac{e^{i\theta} + z}{e^{i\theta} - z} d\rho(\theta)\right\}$$

where ρ is a real measure on the circle, and B_1 and B_2 are Blaschke products. The absolutely continuous part of ρ will be $\frac{1}{2\pi} \log |f| d\theta$. We could also arrange that B_1 and B_2 have no common factor, and then the representation is easily seen to be unique.

Of course, an H^p function is the quotient of two bounded analytic functions. It suffices to show this when f is an outer function in H^p. Then f is the quotient of the bounded outer functions determined by $\log^- |f|$ and $-\log^+ |f|$.

Functions of bounded characteristic are characterized by the property that the integral of $\log^+ |f|$ around the circle of radius r is bounded as $r \to 1$. For this result, we refer the reader to Nevanlinna's book, *Eindeutige Analytischen Funktionen*.

NOTES

The factorization for H^p functions should probably be attributed to Riesz [71], and Herglotz [45]. In the generality of functions of bounded characteristic, the reference is Nevanlinna [64]. A good reference for some of the fundamentals is Beurling's paper [8], in which he coined the terms inner and outer functions. See also Rudin [75], Zygmund [98], Bieberbach [9], Privaloff [70]. The theorem on absolute convergence of Fourier series is in Hardy-Littlewood [41]. See Zygmund [98], as usual, for related questions. For a similar factorization of matrix-valued H^p functions see the papers of Masani [58], Wiener and Masani [59], and Potapov [69].

EXERCISES

1. Prove that

$$\sum_{n=1}^{\infty} \frac{1}{\log n} e^{in\theta}$$

is not the Fourier series of a finite measure on the circle.

2. Use the Hardy-Littlewood theorem to prove that the Taylor series about the origin for $(1 - z)^{1/2}$ is absolutely convergent in the disc.

3. Prove that a bounded analytic function in the right half-plane which vanishes at each positive integer is identically zero.

4. Let f be a function in H^1 and suppose that the functions $z^n f$, $n \geq 0$, span H^1. Prove that f is outer.

5. Let F be an outer function in H^1. Suppose f is in H^1 and f/F is integrable on the unit circle, $|z| = 1$. Prove that f/F is in H^1. Show that this property characterizes outer functions F.

6. Let f be an analytic function in the unit disc. Prove that f is a (constant of modulus 1 times a) Blaschke product if and only if
(a) $|f(z)| \leq 1$
(b) $\lim_{r \to 1} \int_{-\pi}^{\pi} \log |f(re^{i\theta})| d\theta = 0$.

7. Can a Blaschke product be extended continuously from the open unit disc to any point on the circle where its zeros accumulate?

8. Let $|\alpha| < 1$ and let ϕ be the linear fractional map

$$\phi(z) = \frac{z - \alpha}{1 - \bar{\alpha}z}.$$

Then ϕ induces a map $f \to f \circ \phi$ on the space of analytic functions in the disc. Does this map preserve the class H^1? The class of inner functions? The class of outer functions in H^1?

9. Let f be a bounded analytic function in the unit disc, and suppose there is a positive number δ such that

$$|1 - z| + |f(z)| \geq \delta.$$

Prove that there exist *bounded* analytic functions g and h in the disc so that

$$(1 - z)g(z) + f(z)h(z) = 1, \quad |z| < 1.$$

(Hint: Show that one can choose h so that fh extends to be analytic in a neighborhood of $z = 1$ and has the value 1 at $z = 1$.)

10. Let f be an analytic function in the unit disc. Prove that f is the quotient of two bounded analytic functions if and only if

$$\int_{-\pi}^{\pi} \log^+ |f(re^{i\theta})| d\theta$$

is bounded as $r \to 1$. (Hint: Use the $\log^+$ condition to prove the Blaschke product of the zeros of f converges; write $f = Bg$ and get g in the class $\log^+$. Write $g = e^h$ and see what g in $\log^+$ says about h. Details are in Nevanlinna [64].)

11. Let $f \neq 0$ be an analytic function in the unit disc, and suppose f is in H^1. Assume you know nothing about H^1 except its definition. Norm H^1 by

$$\|f\|_1 = \lim_{r \to 1} \frac{1}{2\pi} \int_{-\pi}^{\pi} |f(re^{i\theta})| d\theta$$

or

$$\|f\|_1 = \sup_{r < 1} \frac{1}{2\pi} \int_{-\pi}^{\pi} |f(re^{i\theta})| d\theta.$$

Prove that evaluation at the origin is bounded on H^1, i.e.,

$$|f(0)| \leq K\|f\|_1.$$

Now let $\alpha_1, \alpha_2, \ldots$ be the zeros of f in the open disc which are different from 0 and show that the argument we applied to bounded functions proves the convergence of $\Pi |\alpha_n|$.

12. Let f be a function in H^1. Prove that either of these conditions implies that f is an outer function:
   (a) $1/f$ is in H^1.
   (b) $\operatorname{Re} f(z) > 0$ for $|z| < 1$.

If f is inner, prove that $1 + f$ is outer.

13. Let S be the singular function determined by the positive singular measure μ. Let λ be any point on the unit circle which is in the closed support of μ. Prove that there is a sequence of points z_n in the open disc such that

$$\lim z_n = \lambda \quad \text{and} \quad \lim S(z_n) = 0.$$

CHAPTER 6

ANALYTIC FUNCTIONS WITH CONTINUOUS BOUNDARY VALUES

In this chapter we shall be studying the algebra A of continuous functions on the closed unit disc which are analytic on the open disc. There are various alternative descriptions of A. For example, A is the uniform closure of the polynomials $p(z)$. Or, if we identify each function in A with its boundary values, A consists of the continuous functions on the unit circle whose Fourier coefficients vanish on the negative integers. Beyond the polynomials, the most obvious functions in A are those which have an absolutely convergent Taylor series

$$f(z) = \sum_{n=0}^{\infty} a_n z^n, \quad \sum_{n=0}^{\infty} |a_n| < \infty.$$

Such functions do not exhaust A, since A also contains, e.g., the functions which are sums of uniformly (but not absolutely) convergent Taylor series. We proved in the last chapter that any function in A which is of bounded variation on the unit circle necessarily has an absolutely convergent Taylor series.

The main point of this chapter is to give the complete description of the closed ideals in A. This description will make heavy use of the factorization theory of the last chapter, as well as some new material we shall develop in this chapter. Before we begin to discuss the ideal theory, we want to make a few observations about functions in A.

If $f \neq 0$ is a function in A, we know that the function $\log |f(e^{i\theta})|$ is Lebesgue integrable on the unit circle. In particular, if K is the (closed) set of zeros of f on the circle, then K has Lebesgue measure zero. We shall need a converse for this, namely, if K is an arbitrary closed set of Lebesgue measure zero on $|z| = 1$, then there exists a function in A whose zeros on the closed unit disc are precisely the points of K. This is a result of Fatou, which we shall soon prove; indeed, we shall prove Rudin's generalization of the result which says that, given any continuous complex-valued function g on K, there is a function f in A such that $f = g$ on K. For these constructions we need to know a little about the behavior of conjugate harmonic functions on the boundary.

Conjugate Harmonic Functions

Suppose f is a Lebesgue-integrable function on the circle. For convenience, let f be real-valued. The function

$$h(z) = \frac{1}{2\pi} \int_{-\pi}^{\pi} \frac{e^{i\theta} + z}{e^{i\theta} - z} f(\theta) d\theta$$

is analytic in the open disc. The real part of h is the Poisson integral of f. Thus we know that if $h = u + iv$, the harmonic function u has non-tangential limits which exist and agree with f almost everywhere on the circle. What about the harmonic conjugate v? In other words, if

$$v(r, \theta) = \frac{1}{2\pi} \int_{-\pi}^{\pi} f(\theta - t) Q_r(t) dt$$

where Q_r is the conjugate Poisson kernel,

$$Q_r(t) = \operatorname{Im} \frac{1 + re^{it}}{1 - re^{it}} = \frac{2r \sin t}{1 - 2r \cos t + r^2},$$

does v have non-tangential limits almost everywhere? If so, what are they? Now Q_r is obviously not an approximate identity (positive kernel) like P_r. Nevertheless, the non-tangential limits do exist almost everywhere for v. This is easy to see. It certainly will suffice to prove this when $f \geq 0$. If we assume that $f \geq 0$, the above analytic function h has non-negative real part. Thus, e^{-h} is a bounded analytic function. Therefore, $e^{-(u+iv)}$ has non-tangential limits at almost every point of the circle, and since these limits cannot vanish on a set of positive measure, both u and v have finite non-tangential limits almost everywhere.

This does not answer the question of what the limits are. Since

$$\lim_{r \to 1} Q_r(t) = Q_1(t) = \frac{\sin t}{1 - \cos t} = \operatorname{ctn} \tfrac{1}{2} t$$

one's guess would be that

$$\lim_{r \to 1} v(r, \theta) = \frac{1}{2\pi} \int_{-\pi}^{\pi} f(\theta - t) \operatorname{ctn} \tfrac{1}{2} t \, dt.$$

This is the answer. Since

$$v(r, \theta) = \frac{1}{2\pi} \int_{-\pi}^{\pi} f(\theta - t) Q_r(t) dt$$

$$= -\frac{1}{2\pi} \int_{-\pi}^{\pi} f(\theta + t) Q_r(t) dt$$

$$= -\frac{1}{2\pi} \int_{-\pi}^{\pi} \frac{f(\theta + t) - f(\theta - t)}{2} Q_r(t) dt$$

we are asserting that, as $r \to 1$, $v(r, \theta)$ approaches

$$-\frac{1}{2\pi} \int_{-\pi}^{\pi} \frac{f(\theta + t) - f(\theta - t)}{2 \tan \frac{1}{2}t} dt$$

almost everywhere. In particular, this integral exists for almost every θ. Here we shall not give the whole story, proving only the sufficiency of the existence of this integral.

Theorem. *Let f be an integrable function on the circle and*

$$v(r, \theta) = \frac{1}{2\pi} \int_{-\pi}^{\pi} f(\theta - t) Q_r(t) dt.$$

If θ is any number such that the integral

$$v(\theta) = -\frac{1}{2\pi} \int_{-\pi}^{\pi} \frac{f(\theta + t) - f(\theta - t)}{2 \tan \frac{1}{2}t} dt$$

exists, then $\lim_{r \to 1} v(r, \theta) = v(\theta)$.

Proof. Let

$$\phi_\theta(t) = \frac{f(\theta + t) - f(\theta - t)}{2 \tan \frac{1}{2}t}$$

so that we are assuming ϕ_θ is integrable. Now

$$v(r, \theta) - v(\theta) = \frac{1}{2\pi} \int_{-\pi}^{\pi} \phi_\theta(t) \left[1 - \frac{2r \sin t \tan \frac{1}{2}t}{1 - 2r \cos t + r^2} \right] dt$$

$$= \frac{1}{2\pi} \int_{-\pi}^{\pi} \phi_\theta(t) \frac{(1 - r)^2}{1 - 2r \cos t + r^2} dt.$$

Now if

$$g_r(t) = \frac{(1 - r)^2}{1 - 2r \cos t + r^2}$$

then $0 < g_r(t) < 1$ and $\lim_{r \to 1} g_r(t) = 0$, except at $t = 0$. Since ϕ_θ is integrable, we must have $\int \phi_\theta g_r \to 0$, i.e.,

$$\lim_{r \to 1} v(r, \theta) = v(\theta).$$

Corollary. *If f is differentiable at θ then*

$$\lim_{r \to 1} v(r, \theta) = v(\theta)$$

exists. If, say, f is continuously differentiable on a closed interval $|\theta - \theta_0| \leq \delta$, then on that interval the functions v_r converge uniformly as $r \to 1$.

Proof. The function ϕ_θ is clearly integrable on any interval $|t| \geq \epsilon > 0$. If f is differentiable at θ, then ϕ_θ is bounded as $t \to 0$, so ϕ_θ is integrable. If f is continuously differentiable on $|\theta - \theta_0| \leq \delta$, then we obtain a uniform bound on ϕ_θ for θ in the interval and t small; it is easy to see that $v(r, \theta)$ is uniformly close to $v(\theta)$.

Theorems of Fatou and Rudin

The last corollary enables us to construct special analytic functions with continuous boundary values.

Theorem (Fatou). *Let K be a closed set of Lebesgue measure zero on the unit circle. Then there exists a function in* A *which vanishes precisely on* K.

Proof. Let w be an extended real-valued function on the circle such that

(i) $w = -\infty$ on K, and tends continuously to $-\infty$ as $e^{i\theta}$ approaches K;
(ii) $w \leq -1$ on the circle;
(iii) w is finite-valued and continuously differentiable on $C - K$;
(iv) w is integrable.

Such a w can be found since K has measure zero. One naive way to construct such a function is the following. Since K is closed, the complement C-K is the union of a countable number of disjoint open intervals (arcs) I_n. Let ϵ_n be the length of I_n. Choose a strictly positive and continuously differentiable function y_n on I_n such that $y_n \leq e^{-1}$, y_n tends to zero at the endpoints of I_n, and

$$\int_{I_n} \log y_n \geq -2\epsilon_n.$$

If we define y to be zero on K and $y = y_n$ on I_n, then $0 \leq y \leq e^{-1}$; the zeros of y are precisely the points of K; y is continuous on C and continuously differentiable on C-K; and $\log y$ is integrable. Let $w = \log y$.

Now define

$$h(z) = \frac{1}{2\pi} \int_{-\pi}^{\pi} \frac{e^{i\theta} + z}{e^{i\theta} - z} w(\theta) d\theta.$$

Then h is analytic in the open disc and $\operatorname{Re} h \leq -1$. By property (iii) of w and the Corollary above, h is actually continuous on the complement of K in the closed disc. Since w tends continuously to $-\infty$ at each point of K, the function

$$\operatorname{Re} h(r, \theta) = \frac{1}{2\pi} \int_{-\pi}^{\pi} w(t) P_r(\theta - t) dt$$

tends radially to $-\infty$ for each θ in K.

Now let

$$g = \frac{1}{h}.$$

It is apparent that g is in A, $\operatorname{Re} g \leq 0$, and the zeros of g on the closed disc are exactly the points of K. We remark that $\operatorname{Re} g = 0$ exactly on K.

Theorem (Rudin). *Let* K *be a closed set of Lebesgue measure zero on the unit circle, and let* F *be any continuous complex-valued function on* K. *Then there exists a function in* A *whose restriction to* K *is* F.

Proof. Assume K is non-empty. We can find a function f in A such that
 (i) $f(z) = 1$ if $z \in K$;
 (ii) $|f(z)| < 1$ if z is a point of the closed disc not in K. Just construct g as in the last theorem and let $f = e^g$.

The idea of the proof is this. Using this function f, we prove that the algebra of continuous functions on K which we obtain by restricting A to K is uniformly closed (on K). Then we observe that this algebra of restrictions is (uniformly) dense in the continuous functions on K.

Let h be any function in A. Since $f = 1$ on K and $|f| < 1$ off K,

$$\sup_K |h| = \lim_{n \to \infty} ||f^n h||_\infty$$

where (of course) $||\cdots||_\infty$ denotes the sup norm over the entire closed disc. Each $f^n h$ agrees with h on K. This means that $f^n h = h + g$, where g is a function in A which vanishes on K. We now have

$$\sup_K |h| = \inf_g ||h + g||_\infty, \quad g \in A, \quad g = 0 \text{ on } K.$$

Let S be the subspace of functions in A which vanish on K. Now S is a closed subspace of A and

$$\sup_K |h| = \inf_{g \in S} ||h + g||_\infty.$$

The right-hand number above is the standard quotient norm for the coset $h + S$ of the subspace S. It is easy to verify that, if S is a closed subspace of a Banach space A, the quotient space A/S is complete in the quotient norm. We conclude that the restriction of A to K is uniformly closed.

It is quite easy to see that if K is any proper closed subset of the circle, then the restriction of A to K is dense in the continuous functions on K. Let A_K be the uniform closure on K of the restrictions of the functions in A. By a rotation, we may assume that K omits the open arc $|\theta| < \alpha$ on the circle. If x is a positive number greater than 1, then $(z - x)^{-1}$ is in A_K, indeed, in A. Let x_0 be the infimum of the real numbers $x > -1$ such that $(z - x)^{-1}$ is in A_K. Claim: $x_0 = -1$. Suppose $x_0 > -1$. Then we can find an $x > x_0$ and $\epsilon > 0$ such that $x - x_0 < \epsilon$ but K lies outside the disc $|z - x| < \epsilon$ and $(z - x)^{-1}$ is in A_K. Let $w = (z - x)^{-1}$ so that K lies in the disc $|w| \leq 1/\epsilon$. Now $(z - x_0)^{-1}$ is analytic on $|w| \leq 1/\epsilon$ and is, therefore, uniformly approximable on that disc by polynomials in w. We conclude that $(z - x_0)^{-1}$ is in A_K; indeed, we see by the same argument that $(z - t)^{-1}$ is in A_K for any t satisfying $|x - t| < \epsilon$. Clearly, there is such a real t less than x_0. This contradicts the definition of x_0. We conclude that $x_0 = -1$; in particular, $1/\bar{z} = z$ is in A_K. By Fejer's theorem, A_K contains every continuous function on K.

We remark that the second half of the above proof (that is, that on a proper closed subset of the circle every continuous function is a uniform limit of polynomials) is a very old theorem which has a variety of proofs. The one we gave is one of the more elementary ones and is modeled after the proof of Runge's approximation theorem. Another interesting, but less elementary, proof can be based on the theorem of F. and M. Riesz on analytic measures. Suppose K is a proper closed subset of the circle; let A_K be the uniform closure on K of the polynomials. Suppose $A_K \neq C(K)$. By the Hahn-Banach theorem there is a bounded linear functional on $C(K)$ which annihilates A_K but is not zero. In other words, there is a non-zero finite complex measure μ on K such that $\int p d\mu = 0$ for every polynomial p. According to F. and M. Riesz, any such measure on the circle has the form $d\mu = f d\theta$ with f in H^1. Since μ is supported on K, we see that f vanishes on an open arc of the circle. So $f = 0$ and $\mu = 0$, which is a contradiction.

We should also point out that Rudin proved more than we have stated. He proved that if F is continuous on the closed set of measure zero, K, then there is an f in A such that $f = F$ on K and f is bounded on the disc by the maximum of F on K.

Bishop has generalized the Rudin theorem as follows. Let A be a uniformly closed algebra of continuous complex-valued functions on a compact Hausdorff space X ($1 \in A$). Let K be any closed subset of X with this property: if μ is a complex measure on X which is orthogonal to A, the total variation of μ on K is zero. Then the restriction of A to K is $C(K)$ (without closing).

The Closed Ideals of A

As we said earlier, we wish to describe the closed ideals in A. If R is any ring, it is obviously of considerable interest to describe the ideals in R. Needless to say, this is usually extremely difficult. If R is a commutative ring with identity, one usually says that he knows the "ideal theory" of R if he can answer such basic questions as the following:

(1) What are the maximal ideals of R?
(2) Is every proper ideal an intersection of maximal ideals?

(3) What are the primary ideals of R, i.e., those contained in precisely one maximal ideal?

(4) Is every ideal an intersection of primary ideals?

In a topological ring (or Banach algebra) such as our algebra A, one normally restricts his attention to closed ideals. The task of describing just the closed ideals is usually insurmountable; however, for our algebra A they can all be described very precisely. This ideal theory was first done by Beurling, but he never published the results. Later, Rudin independently obtained the results (*Canadian Journal of Mathematics*, 1957).

Since A happens to be a commutative Banach algebra with a single generator, it is quite easy to find the maximal ideals of A, if one is willing to use the general result that, for any maximal ideal in a commutative Banach algebra with identity, the associated quotient field is the field of complex numbers. We shall comment on this later, but for now it is of no particular aid to us, since the description of the maximal ideals will soon drop out of the general assault on the ideal theory.

If λ is a point in the closed unit disc, it is apparent that

$$\{f \in A\,; f(\lambda) = 0\}$$

is a maximal ideal in A. We shall soon see that there are no others. What other closed ideals in A are there? Certainly $z^2 A$, the set of functions in A such that $f(0) = f'(0) = 0$ is a closed ideal. It is also primary, since it is contained in the single maximal ideal $\{f; f(0) = 0\}$. Thus, not every closed ideal in A is an intersection of maximal ideals. There is an obvious extension of this example. Choose a sequence of points α_n in the open disc and a sequence of non-negative integers p_n, and let J be the set of functions in A which have a zero of order at least p_n at α_n. Clearly, J is a closed ideal. Of course, we may have $J = 0$. But if the α_n approach the boundary of the disc rapidly, if the p_n are not too large, and if the α_n do not cluster on a set of positive measure on the circle, we shall have $J \neq 0$. For example, if $\alpha_n = 1 - 2^{-n}$ and $p_n = n$, we have $J \neq 0$.

These are the obvious closed ideals in A: those determined by prescribing orders of zeros at points in the open disc. From what we have already done we can see that there are others. Let K be a closed set of measure zero on the circle, and let J_K be the set of functions in A which vanish on K. We know that J_K is a non-zero ideal (which is obviously closed). There is another type of ideal which is more subtle. This is one determined by the rate at which the functions in it tend to zero as z approaches the boundary. A simple example of this is the following. Let

$$F(z) = \exp\left[\frac{z+1}{z-1}\right].$$

Then F is not in A, but is an inner function. Indeed, F is the singular function determined by the unit point mass at $z = 1$. Let M be the (maximal) ideal of all functions in A which vanish at $z = 1$, and let $J = FM$. Then J is a closed ideal in A. For if g is in A and $g(1) = 0$, then Fg is in A, because F is analytic and of modulus 1 at every point of the circle other than $z = 1$. Of course J is closed, because M is closed and F is of modulus 1 almost everywhere. Furthermore, $J \neq M$ because $(1 - z)$ is not in J. If it were, $(1 - z)F^{-1}$ would be bounded, which it is not.

Now we make a much more general type of closed ideal. Subsequently we show that this is the only type.

Theorem. *Let* K *be a closed set of Lebesgue measure zero on the unit circle. Let* F *be an inner function such that*

(i) *if* $\alpha_1, \alpha_2, \ldots$ *are the zeros of* F *in the open disc, then every accumulation point of the* α_n *is in* K;

(ii) *the measure determining the singular part of* F *is supported on* K.

Let J *be the set of functions of the form* Fg, *where* g *is a function in* A *which vanishes on* K. *Then* J *is a closed non-zero ideal in* A.

Proof. Let $F = BS$, where B is the Blaschke product with zeros $\alpha_1, \alpha_2, \ldots$ and S is the inner function determined by the positive singular measure μ. Every accumulation point of the α_n is in K. Thus, B is analytic in the complement of K in the complex plane. The closed support of μ is contained in K. Thus, S is analytic on the complement of K. If g is any function in A which vanishes on K, the fact that F is bounded makes it obvious that Fg is in A. If J is the set of such functions Fg, certainly J is an ideal in A. Also, J is non-zero, because the zero measure of K guarantees the existence of a $g \neq 0$ in A which vanishes on K. To see that J is closed, argue as follows. Let $\{Fg_n\}$ be a sequence of functions in J which converges to a function f in A. Then

$$\|Fg_n - f\|_\infty \to 0,$$

and, since F is of modulus 1 on the circle,

$$\|g_n - \bar{F}f\|_\infty \to 0$$

(essential sup on the circle). Thus, $\{g_n\}$ converges uniformly to $\bar{F}f$ on the circle. So $\bar{F}f = F^{-1}f = g$, where $g = \lim g_n$. Clearly, f is in J (i.e., $f = Fg$).

Now we wish to show that every non-zero closed ideal in A has the above form. To prove this, begin with such an ideal J and let K be the closed set of measure zero on the circle obtained by intersecting the zeros on the circle of all the functions in J. How do we produce the inner function F? F will have to be an inner function which divides every f in J. Consequently, F must divide the inner part of every f in J. Indeed, one feels that F should be the "greatest common divisor" of these inner parts. We need the following.

If F and G are inner functions, we shall say that F **divides** G if G/F is a bounded analytic function. When F divides G, it is clear that G/F is, again an inner function.

Lemma. *Let $\mathfrak{F}$ be any non-empty family of inner functions. Then there is a unique inner function* F *with these two properties:*
  (i) F *divides every function in* $\mathfrak{F}$.
  (ii) *If* F_1 *is an inner function which divides each function in* $\mathfrak{F}$, *then* F_1 *divides* F.

Proof. Let $F = BS$ and $F_1 = B_1 S_1$ be inner functions. If F divides F_1, then clearly B divides B_1, i.e., every zero of F in the open disc (with multiplicity) is also a zero of F_1. If F divides F_1, then clearly S must divide S_1. Now

$$\frac{S_1}{S}(z) = \exp\left[\int \frac{e^{i\theta} + z}{e^{i\theta} - z}(d\mu - d\mu_1)\right]$$

where μ is the positive singular measure defining S and μ_1 is the corresponding measure for S_1. Now S_1/S is bounded if and only if $\mu - \mu_1 \leq 0$, i.e., if and only if $\mu_1 \geq \mu$.

Given the family $\mathfrak{F}$, the function F we are seeking is defined as follows. The Blaschke product for F is the one formed from the common zeros of the functions in $\mathfrak{F}$. The measure μ for F will be the "greatest" positive measure on the circle which is dominated by the corresponding measures of all functions in $\mathfrak{F}$. If there is such a measure, we clearly have the F we want.

All we need demonstrate is that any family of positive measures has a greatest lower bound. This is easy to verify. If $\{\mu_\alpha\}$ is such a family of measures, the greatest lower bound of the μ_α is the measure μ defined by

$$\mu(E) = \inf_{P} \sum_{j=1}^{n} \inf_{\alpha} \mu_\alpha(E_j)$$

where P ranges over all partitions of the set E into the disjoint union of Baire sets $E_1, \ldots, E_n$. Upon checking that μ is a measure, we are done.

Needless to say, we call F the **greatest common divisor** of the inner functions in the family $\mathfrak{F}$.

Theorem. *Let J be a non-zero closed ideal in* A, *and let K be the intersection of the zeros of the functions in J on the unit circle. Let F be the greatest common divisor of the inner parts of the non-zero functions in J. Then J is precisely the set of functions of the form* Fg, *where g ranges over the functions in* A *which vanish on* K.

Proof. Virtually by definition, every function f in J has the form Fg, where g is in A and vanishes on K. Just let g be the quotient of the inner part of f by F, multiplied by the outer part of f.

If $\tilde{J}$ denotes the set of functions in A of the form f/F with f in J, then clearly $\tilde{J}$ is a closed ideal in A which is contained in the ideal of all functions vanishing on K. We are to prove that these two ideals are identical. What this tells us is that we need only prove the theorem for the case $F = 1$.

Assume that we have a closed ideal J in A, such that the greatest common divisor of the inner parts of its non-zero elements is 1. Let K be the common zeros of the functions in J on the circle, and let $I = I(K)$ be the ideal of all functions in A which vanish on K. We shall prove that $J = I(K)$. To do this, it will suffice to prove that any complex measure on the circle which is orthogonal to J is also orthogonal to I. This says that, as closed subspaces of the continuous functions on the circle, J and I are annihilated by exactly the same bounded linear functionals.

Let μ be a finite complex Baire measure on the circle such that μ is orthogonal to J:

$$\int f d\mu = 0, \quad f \in J.$$

Fixing f in J, we have (since J is an ideal)

$$\int z^n f d\mu = 0, \quad n = 0, 1, 2, \ldots .$$

This means that $f d\mu$ is an "analytic" measure. The theorem of F. and M. Riesz tells us that

$$f d\mu = \frac{1}{2\pi} H_f d\theta$$

where H_f is in H^1. Note that H_f also vanishes at the origin.

Let

$$d\mu = \frac{1}{2\pi} \phi d\theta + d\mu_s$$

be the Lebesgue decomposition for μ (ϕ in L^1 and μ_s singular). From above, we see that $f d\mu_s = 0$ for every f in J. Since the functions in J have no common zero off K, this shows that μ_s is supported on K. We also have associated with each f in J an H^1 function H_f such that $f\phi = H_f$ almost everywhere. If $f \neq 0$, this just says that ϕ agrees almost everywhere with the non-tangential limits of the meromorphic function H_f/f on the disc. Among other things, we see that the meromorphic function H_f/f (for $f \neq 0$ in J) is independent of f. If f, g are in J and non-zero, we have $H_f/f = H_g/g$ almost everywhere on the circle; hence,

$$gH_f - fH_g = 0 \quad \text{a.e. on the circle.}$$

But $gH_f - fH_g$ is in H^1. Thus, this function vanishes identically on the disc:

$$\frac{H_f(z)}{f(z)} = \frac{H_g(z)}{g(z)}.$$

Let us call this meromorphic function M. Note that M is actually analytic, for we are assuming that the g.c.d. of the inner parts of the functions in J is 1. At any point λ in the open disc we have an f in J for which $f(\lambda) \neq 0$. Since $M = H_f/f$, we see that M is analytic at λ.

We wish to prove that M is in H^1. Since the non-tangential limits of M agree with ϕ almost everywhere, this will show us that ϕ is in H^1. Of course, M may be 0, but that case requires no comment. If we choose a particular $f \neq 0$ in J (or use the fact that M is of bounded characteristic) and remember that M is analytic, we see that

$$M(z) = \lambda B(z) \exp\left[\int \frac{e^{i\theta} + z}{e^{i\theta} - z} d\rho(\theta)\right]$$

where B is a Blaschke product and ρ is the real measure defined by

$$d\rho = \frac{1}{2\pi} \log |M| d\theta + d\rho_1 - d\rho_2$$

$$= \frac{1}{2\pi} \log |\phi| d\theta + d\rho_1 - d\rho_2$$

ρ_1 and ρ_2 being positive singular measures on the circle. For any f in J we have fM in H^1. This clearly means that, if μ_f is the measure defining the singular part of f, $\mu_f \geq \rho_1$. But the inner factors of the f's have g.c.d. 1, so $\rho_1 = 0$. Thus

$$M(z) = \lambda B(z) \exp\left[-\int \frac{e^{i\theta} + z}{e^{i\theta} - z} d\rho_2\right] \exp\left[\frac{1}{2\pi} \int_{-\pi}^{\pi} \frac{e^{i\theta} + z}{e^{i\theta} - z} \log |\phi| d\theta\right]$$

where ϕ is in L^1, $\log |\phi|$ is in L^1, and ρ_2 is a positive singular measure. Thus M is in H^1; i.e., ϕ is in H^1 of the circle. Of course, M also vanishes at the origin, because each H_f does.

Now let h be any function in A which vanishes on K. Then

$$\int h d\mu = \frac{1}{2\pi} \int h\phi d\theta + \int h d\mu_s.$$

Since ϕ is in H^1 and vanishes at the origin, the first integral is zero. Since μ_s is supported on K, and since h vanishes on K, the second integral is zero. That completes the proof.

Corollary. *Every maximal ideal of* A *is of the form*

$$M_\lambda = \{f \in A; f(\lambda) = 0\}$$

for some point λ *in the closed unit disc.*

Proof. First, we need to observe that if J is any proper ideal in A, the closure of J is a proper closed ideal in A. The closure of J is clearly an ideal; we need only show that it is proper. If f belongs to J, we must have $||1 - f||_\infty \geq 1$. If $||1 - f||_\infty < 1$, then obviously $1/f$ is in A, and no element of a proper ideal can be invertible. Thus, $||1 - f||_\infty \geq 1$ will also hold for every f in the closure of J. If M is a maximal ideal, the above remarks make it clear that M is closed. The description we have given for closed ideals makes it apparent that M must be an M_λ.

Corollary. *If* $f_1, \ldots, f_n$ *are functions in* A *which have no common zero on the closed disc, then there exist functions* $g_1, \ldots, g_n$ *in* A *such that*
$$f_1 g_1 + \cdots + f_n g_n = 1.$$

Proof. The set of all functions of the form $f_1 g_1 + \cdots + f_n g_n$ is an ideal. If it does not contain 1, it is a proper ideal and must (by Zorn's lemma) be contained in a maximal ideal.

Corollary. *Every closed ideal in* A *is the principal closed ideal generated by a function in* A.

Proof. Certainly the zero ideal is principal. If J is a non-trivial closed ideal, let K and F be as in the theorem. Let g be any function in A whose zeros are precisely K and let f be the outer part of g. Then it is clear that Ff generates J.

Corollary. *The closed ideals* J *in* A *which are primary, i.e., contained in precisely one maximal ideal, are those of the following types:*

(i) *J is the principal ideal generated by* $(z - \alpha)^k$, *where* k *is a positive integer and* α *is a point of the open unit disc.*

(ii) *J is the (closed) principal ideal generated by*
$$f(z) = (z - \lambda) \exp\left[\rho \frac{z + \lambda}{z - \lambda}\right]$$
where $|\lambda| = 1$ *and* ρ *is a non-negative real number.*

Proof. If J is primary and contained in the maximal ideal M_α with $|\alpha| < 1$, then the corresponding set K is empty and the corresponding inner function must be
$$F(z) = \left[\frac{\bar\alpha}{|\alpha|} \frac{\alpha - z}{1 - \bar\alpha z}\right]^k.$$

If J is primary and in M_λ with $|\lambda| = 1$, then $K = \{\lambda\}$ and the corresponding F must be
$$F(z) = \exp\left[-\int \frac{e^{i\theta} + z}{e^{i\theta} - z} d\mu(\theta)\right]$$
where μ is a positive measure concentrated at the point λ. If we let $\rho = \mu(\{\lambda\})$, we have (ii).

It is not true that every closed ideal is an intersection of primary ideals. If J, K, F are as above, it is easy to verify that J is an intersection of primary ideals if and only if the measure μ which determines the singular part of F is discrete, i.e., the sum of a countable number of point masses.

Let us review. Suppose we have a closed ideal J in A, $0 \neq J \neq A$.

Then J will be contained in a certain set of maximal ideals. This set is usually called the **hull** of J. Thus the hull of J is the set of common zeros on the disc of the functions in J. Now hull (J) will look like this:

  (i) hull (J) is a non-empty proper closed subset of $|z| \leq 1$.

  (ii) There is either a finite or a countably infinite number of points of the open disc which belong to hull (J). If there is an infinite number of such points $\alpha_1, \alpha_2, \ldots$, then

$$\sum_{n=1}^{\infty} (1 - |\alpha_n|) < \infty.$$

  (iii) The intersection of hull (J) with the unit circle is a closed set K of Lebesgue measure zero, and K contains every accumulation point of the points α_n of (ii).

Of course, any closed set satisfying (i), (ii), and (iii) is the hull of some closed ideal of A, namely, the ideal of all functions in A which vanish on that set. This ideal is simply the intersection of the maximal ideals which contain J, for any J having that hull. This intersection is usually called the **kernel** of the hull of J.

If H is a hull, we have a complete description of all the closed ideals J which have H as their hull. Each such J is obtained as follows. For each α_n in H select a positive integer p_n. The only constraint on the choice of the p_n is that $\sum p_n(1 - |\alpha_n|) < \infty$. Choose a finite positive measure μ on $H \cap C$. Let J be the set of functions f in A such that

  (a) f vanishes on H;
  (b) f has a zero of order at least p_n at α_n;
  (c) the function

$$f(z) \exp \left[\int \frac{e^{i\theta} + z}{e^{i\theta} - z} d\mu(\theta) \right]$$

is bounded as $|z| \to 1$. For each such choice of the p_n and μ we obtain a closed ideal J and distinct choices give distinct ideals. The p_n specify orders of zeros inside the disc, and μ specifies "order" of vanishing at the boundary.

We already looked at the case when H is a single point. If $H = \{\lambda\}$ with $|\lambda| < 1$, we obtain a countable number of closed ideals with hull H. If $|\lambda| = 1$, then we have a continuum of distinct closed ideals with hull $H = \{\lambda\}$.

Commutative Banach Algebras

We wish to prove Wermer's theorem that A is a maximal closed subalgebra of the continuous functions on the unit circle. For this we shall not need the ideal theory which we have just completed; however, we do need the characterization of the maximal ideals of A (which we have). As we mentioned earlier, this characterization of the maximal ideals is easy to obtain if one knows one fundamental theorem on commutative Banach algebras, and we shall now derive the result from that theorem, since the concepts involved are relevant to the proof of Wermer's theorem.

A **commutative Banach algebra** is a commutative complex linear algebra B, which is equipped with a norm under which it is a Banach space and for which the norm is related to the multiplication by

$$||xy|| \leq ||x||\, ||y||.$$

If B has an identity for multiplication we also require that the identity has norm 1. Certainly A is such an object, i.e., a commutative Banach algebra with identity (using the sup norm). For another example, take

$$B = H^\infty, \quad \text{with} \quad ||f|| = \sup |f(z)|, \quad |z| < 1.$$

Let B be a commutative Banach algebra with identity. We denote the identity of B by 1, and abbreviate $\lambda 1$ to λ.

Lemma 1. *If $||1 - x|| < 1$, then x is invertible.*

Proof. Since $||y^n|| \leq ||y||^n$, the series $1 + (1-x) + (1-x)^2 + \cdots$ converges in B to $x^{-1} = [1 - (1-x)]^{-1}$.

Lemma 2. *If $|\lambda| > ||x||$, then $(x - \lambda)$ is invertible.*

Proof. $(x - \lambda) = \lambda\left(\dfrac{1}{\lambda}x - 1\right)$. It suffices to invert $1 - \dfrac{1}{\lambda}x$. But

$$\left\|\frac{1}{\lambda}x\right\| = \frac{1}{|\lambda|}||x|| < 1$$

so $1 - \dfrac{1}{\lambda}x$ is invertible by Lemma 1.

Lemma 3. *The set of invertible elements of* B *is open, and on that set the map* $x \to x^{-1}$ *is continuous.*

Proof. Suppose x is invertible and let y be any element of B such that

$$||x - y|| < ||x^{-1}||^{-1}.$$

Then

$$||1 - x^{-1}y|| = ||x^{-1}(x - y)||$$
$$\leq ||x^{-1}||\, ||x - y|| < 1.$$

Thus $x^{-1}y$ is invertible by Lemma 1; so y is invertible. This shows that the set of invertible elements is open. If we use the geometric series for the inverse of $(x^{-1}y)$, it is easy to verify that

$$||x^{-1} - y^{-1}|| < ||x - y||\, ||x^{-1}||(||x^{-1}||^{-1} - ||x - y||)$$

which shows that inversion is continuous.

Analytic Functions with Continuous Boundary Values

Lemma 4. *The set of complex numbers λ such that $(x - \lambda)$ is invertible is an open set in the complex plane, the complement of which is compact and non-empty.*

Proof. The set of such λ is open by Lemma 3. Its complement is called the **spectrum** of x. It is compact because (by Lemma 2) it is contained in the closed disc of radius $||x||$. To prove that this spectrum is non-empty, we argue as follows. Let F be any bounded linear functional on B. Define a complex-valued function f on the complement of the spectrum of x by

$$f(\lambda) = F[(x - \lambda)^{-1}].$$

Now f is analytic. For

$$\frac{f(\lambda + h) - f(\lambda)}{h} = \frac{1}{h}[F((x - \lambda - h)^{-1}) - F((x - \lambda)^{-1})]$$

$$= \frac{1}{h} F[(x - \lambda - h)^{-1} - (x - \lambda)^{-1}]$$

$$= \frac{1}{h} F[h(x - \lambda - h)^{-1}(x - \lambda)^{-1}]$$

$$= F[(x - \lambda - h)^{-1}(x - \lambda)^{-1}].$$

If we let $h \to 0$, we see (by the continuity of inversion and F) that f is differentiable and

$$f'(\lambda) = F[(x - \lambda)^{-2}].$$

Note that $\lim_{|\lambda| \to \infty} f(\lambda) = 0$, because

$$f(\lambda) = \frac{1}{\lambda} F\left[\left(\frac{1}{\lambda} x - 1\right)^{-1}\right],$$

and as $|\lambda| \to \infty$, $\frac{1}{\lambda} x \to 0$, and (since inversion and F are continuous)

$$F\left[\left(\frac{1}{\lambda} x - 1\right)^{-1}\right] \to -F(1).$$

If the spectrum of x is empty, then for each bounded linear functional F the associated f is an entire function which tends to zero at infinity. By Liouville's theorem $f(\lambda) \equiv 0$; in particular, $0 = f(0) = F(x^{-1})$. So x^{-1} is killed by every bounded linear functional on B. Thus $x^{-1} = 0$, a complete absurdity.

Lemma 5. *A commutative Banach algebra which is a field is (isomorphic to) the field of complex numbers.*

Proof. Suppose B is a field. Let x be an element of B. By Lemma 4 there is a scalar λ such that $x - \lambda$ is not invertible. But B is a field, so $x - \lambda = 0$. Thus, every element of B is a scalar multiple of the identity.

Theorem. *Let B be a commutative Banach algebra with identity, and let M be a maximal ideal in B. The quotient algebra B/M is isometrically isomorphic to the field of complex numbers. Thus M is the kernel of a homomorphism of B onto the field of complex numbers. Also, M is closed and this homomorphism is continuous.*

Proof. Now $||1 - x|| \geq 1$ for every x in M. Otherwise, M would contain an invertible element and not be a proper ideal. The same inequality holds for all x in the closure of M. This closure is, therefore, a proper ideal containing M and must be equal to M.

The quotient space B/M inherits a natural quotient norm

$$||x + M|| = \inf ||x + y||, \quad y \in M.$$

With this norm B/M is a Banach space. Also, B/M is a commutative linear algebra. It is easy to verify that $(1 + M)$ has norm 1 and

$$||(x + M)(y + M)|| \leq ||x + M||\,||y + M||;$$

hence B/M is a commutative Banach algebra with identity. Since M is a maximal ideal, B/M is a field. By Lemma 5 we see that B/M is isomorphic to the complex numbers. The quotient map from B to B/M may now be regarded as a complex homomorphism of B. It is not only continuous, but norm-decreasing. That completes the proof.

If B is a commutative Banach algebra with identity, and if ϕ is a homomorphism of B onto the field of complex numbers, the kernel of ϕ is a maximal ideal in B. By the last theorem, this kernel is closed and ϕ is automatically continuous; indeed, $|\phi(x)| \leq ||x||$. The last theorem really tells us that there is a one-one correspondence between the complex homomorphisms of B and the maximal ideals of B. Another way to say this is that we may identify the maximal ideals of B with those (necessarily bounded) linear functionals $\phi \neq 0$ on B which happen to be multiplicative: $\phi(xy) = \phi(x)\phi(y)$.

This theorem certainly settles the question of the maximal ideals in our algebra A of continuous functions on the closed disc which are analytic in the interior. Suppose M is a maximal ideal in A. Then M is the kernel of a homomorphism ϕ from A onto the complex numbers. This ϕ is necessarily continuous; in fact,

$$|\phi(f)| \leq ||f||_\infty.$$

Let $\lambda = \phi(z)$. Then $|\lambda| \leq 1$. This determines ϕ on the polynomials

$$\phi(\Sigma\, a_n z^n) = \Sigma\, a_n \lambda^n;$$

that is, ϕ evaluates every polynomial at λ. These polynomials are dense in A, and since ϕ is continuous, ϕ must just be "evaluation at λ." In particular, M is the set of functions in A which vanish at λ.

Wermer's Maximality Theorem

We now turn to Wermer's maximality theorem. For the most part, we shall regard A as a uniformly closed algebra of functions on the unit circle.

Theorem (Wermer). *A is a maximal closed subalgebra of the continuous complex-valued functions on the unit circle. In other words, if f is a continuous complex-valued function on the circle which is not in A, then polynomials in z and f are dense in the continuous functions on the circle.*

Proof. Let C denote the algebra of all continuous complex-valued functions on the circle. Suppose B is a uniformly closed subalgebra of C which contains A. We shall prove that either $B = A$ or $B = C$. This will be done by considering the complex homomorphism ϕ of A obtained by evaluating at the origin:

$$\phi(f) = f(0) = \frac{1}{2\pi}\int_{-\pi}^{\pi} f(e^{i\theta})d\theta.$$

First, suppose that ϕ does not extend to a complex homomorphism of B, i.e., that there is no complex homomorphism of B whose restriction to A is ϕ. Then there is no non-zero complex homomorphism of B which sends the function z into 0, because ϕ is the only homomorphism of A with this property. This means that z lies in no maximal ideal of B and, hence, that z lies in no proper ideal of B. Therefore, $zB = B$, so $1/z = \bar{z}$ is in B. By Fejer's theorem $B = C$.

Suppose that there is a complex homomorphism $\tilde{\phi}$ of B whose restriction to A is ϕ. In particular, $\tilde{\phi}$ is a linear functional on B of bound 1, and can (by the Hahn-Banach theorem) be extended to a linear functional of norm 1 on C. There is thus a finite complex measure μ on the circle such that

$$\tilde{\phi}(f) = \int f d\mu$$

for all f in B, and such that the total variation of μ is 1. Since $\tilde{\phi}(1) = 1$, we have $\int d\mu = 1$. Now it is easy to see that a complex measure which has total variation 1 and integral 1 must be a positive measure. Since B contains A,

$$\int e^{in\theta} d\mu(\theta) = \tilde{\phi}(z^n)$$
$$= \phi(z^n)$$
$$= z^n(0)$$
$$= 0$$

for $n = 1, 2, 3, \ldots$. Since μ is a real measure, this implies $d\mu = \frac{1}{2\pi} d\theta$.

Now let f be any function in B. Then

$$\frac{1}{2\pi}\int_{-\pi}^{\pi} e^{in\theta} f(e^{i\theta}) d\theta = \int z^n f d\mu$$
$$= \tilde{\phi}(z^n f)$$
$$= \tilde{\phi}(z^n) \tilde{\phi}(f)$$
$$= z^n(0) \tilde{\phi}(f)$$
$$= 0$$

for $n = 1, 2, \ldots$. Thus f is in A. We conclude that $B = A$.

Wermer's first proof of this theorem, although not lengthy, made use of more classical analysis than does the argument we have presented (chiefly, some of the boundary value theorems for H^p). The proof above is due to Singer and Hoffman. In reading the proof above, Paul Cohen extracted from it a very elementary proof which is worth presenting. Suppose B is a closed subalgebra of C which contains A properly. Clearly, then, we can find a function f in B whose (-1) Fourier coefficient is 1. By Fejer's theorem we can find polynomials p and q such that

$$zf = 1 + zp + \bar{z}\bar{q} + h$$

where h is a continuous function of sup norm less than $\frac{1}{2}$. Choose a positive number $M \geq \|zq - \bar{z}\bar{q}\|_\infty$. For any $\delta > 0$

$$\|1 + \delta(zq - \bar{z}\bar{q})\|_\infty \leq 1 + \delta^2 M^2$$

because $zq - \bar{z}\bar{q}$ is pure imaginary. Now

$$\delta \bar{z}\bar{q} = \delta(zf - 1 - zp) - \delta h$$
$$= zg - \delta h - \delta$$

where g is in B. Since $|h| < \frac{1}{2}$, we have

$$\|1 + \delta + z(g + \delta q)\|_\infty \leq 1 + \delta^2 M^2 + \frac{\delta}{2}.$$

If we choose δ so that $\delta < 1/2M^2$, we have

$$\|1 + \delta + z(g + \delta q)\|_\infty < 1 + \delta.$$

Now g, q, and z are in B, so the inequality above shows that $z(g + \delta q)$ is invertible in B. But then z is invertible in B, i.e., $\bar{z}$ is in B. Thus $B = C$ by Fejer's theorem.

Corollary. *If K is a proper closed subset of the unit circle, then every continuous complex-valued function on K is a uniform limit of polynomials.*

Proof. Of course we have already given two proofs of this result, but this one is also interesting. Let B be the set of all continuous functions f on the circle such that the restriction of f to K is uniformly approximable on K by polynomials. Clearly, B is a uniformly closed algebra of functions on the circle which contains A. Since K is a proper closed subset of the circle, and since B contains (in particular) every continuous function which vanishes on K, we certainly cannot have $B = A$. By the maximality of A we see that B contains all continuous functions on the circle.

NOTES

Beurling never published the ideal theory for A. The ideas are in Beurling [8]. Rudin's paper on the ideal theory is Rudin [75]; his paper [74] contains the extension of continuous functions from a set of measure zero. This was also done by Carleson [17]. See Bishop [11] for a generalization. The original work on non-trivial extensions of the zero function is Fatou [27]. The behavior of conjugate harmonic functions on the boundary is studied in some detail in Zygmund [98]. For a discussion of commutative Banach algebras, see Gelfand [32] or Gelfand-Raikov-Šilov [33]. Wermer's maximality theorem is Wermer [90]. The proof given here is in Hoffman-Singer [49]. See Cohen [20] for his proof. Maximal algebras of continuous functions are discussed in detail in Hoffman-Singer [49]. See Bishop [10] for a direct generalization of Wermer's theorem: On a simply connected compact set in the plane which has an interior, the uniform closure of the polynomials is a maximal closed subalgebra of the continuous functions on the boundary. Maximality is a special type of approximation theorem. For more on approximation by analytic functions, see the books of Walsh [89], Ahiezer [1], the papers of Mergelyan [60, 61], Szasz [85], and Müntz [62].

EXERCISES

1. Show that $f(\theta) = (1 - e^{i\theta}) \exp\left[\dfrac{1 + e^{i\theta}}{1 - e^{i\theta}}\right]$ is a continuous function on the unit circle, that there is an analytic function in the disc with non-tangential limits almost everywhere equal to f, but that f is not a uniform limit of polynomials.

2. If f is in H^∞ and f^2 is in A, does it follow that f is in A ($A =$ uniform closure of polynomials)? What if we assume only that the boundary values of f^2 are continuous? Answer both these questions, i.e., f^2 is continuous on the closed disc or has continuous boundary values, when f is a bounded harmonic function.

3. If f is a continuously differentiable function on the unit circle and

$$g(x) = \frac{1}{2\pi} \int_{-\pi}^{\pi} \frac{f(x + t) - f(x - t)}{2 \tan \frac{1}{2} t} \, dt,$$

evaluate
$$\frac{1}{2\pi} \int_{-\pi}^{\pi} \frac{g(x + t) - g(x - t)}{2 \tan \frac{1}{2} t} \, dt.$$

4. Is every f in H^∞ the derivative of a function in A? What about every f in H^1?

5. Prove the approximation theorem of Runge [78]: If K is a compact set in the plane with a connected complement, then every function analytic on a neighborhood of K can be uniformly approximated on K by polynomials. Outline of proof: (i) Use the argument we employed for K on the circle, to prove that $1/(z - \alpha)$ is so approximable for every α not in K. (ii) Given f analytic in an open set U about K, choose a rectifiable closed path Γ in U which winds around each point of K exactly once. Write f in terms of its Cauchy integral representation on Γ, and observe that the approximating sums for the integral give a uniform approximation to f by rational functions.

6. Let $\{D_n\}$ be a sequence of closed discs centered at the origin, so that D_n lies in the interior of D_{n+1} and the union of the D_n is the open unit disc. In D_n choose a compact set K_n contained in the interior of D_n and not meeting D_{n-1}. Suppose K_n has a connected complement in the plane. Now use Runge's theorem (Exercise 5) to prove that any function f which is defined and analytic on a neighborhood of the union of the K_n can be uniformly approximated on that union by functions analytic in the open unit disc. Outline of proof: Given f and $\epsilon > 0$, approximate f on D_1 by a polynomial p_1, uniformly within $\epsilon/2$. Choose a polynomial p_2 which is uniformly within $\epsilon/4$ of p_1 on D_1 and uniformly with $\epsilon/4$ of f on K_2. Then approximate p_2 on D_2 and f on K_3 by a p_3, within $\epsilon/8$, etc. The sequence $\{p_n\}$ converges to a function approximating f.

7. Use the result of Exercise 6 to construct

(a) an analytic function in the unit disc which has a radial limit at no point of the unit circle;

(b) (Remmert) three analytic functions f, g, and h in the disc which separate the points of the disc; at each point one of them has a non-vanishing derivative; and $|f| + |g| + |h|$ tends uniformly to $+\infty$ at the boundary [take $h(z) = (1 - z)^{-1}$ and obtain f and g from Exercise 6, using sequences of annuli notched near the positive axis];

(c) a sequence of analytic functions in the disc which converges pointwise to zero, but does not converge uniformly on compact sets.

8. For which functions f in A (the uniform closure of the polynomials) does f belong to the closed ideal generated by f^2?

9. Let $f_1, \ldots, f_n$ be analytic functions in the open unit disc which have no common zero in that open disc. Prove that there exist analytic functions $g_1, \ldots, g_n$ in $|z| < 1$ such that $f_1 g_1 + \cdots + f_n g_n = 1$. (Hint: Let D_k be the closed disc of radius $1 - 1/(k+1)$. Use the corresponding result for analytic functions with continuous boundary values to find $g_1^{(1)}, \ldots, g_n^{(1)}$ continuous on D_1 and analytic in the interior of D_1 such that $\Sigma f_j g_j^{(1)} = 1$ on D_1. Now prove there are $g_1^{(2)}, \ldots, g_n^{(2)}$ continuous on D_2 and analytic in its interior such that

$$\Sigma f_j g_j^{(2)} = 1 \text{ on } D_2 \quad \text{and} \quad |g_j^{(1)} - g_j^{(2)}| < 1 \text{ on } D_1.$$

Get $g_1^{(3)}, \ldots, g_n^{(3)}$ such that

$$\Sigma f_j g_j^{(3)} = 1 \text{ on } D_3 \quad \text{and} \quad |g_j^{(2)} - g_j^{(3)}| < \tfrac{1}{2} \text{ on } D_2, \text{ etc.}$$

10. Let $H(D)$ be the full ring of analytic functions on some open set D in the complex plane. Prove that every homomorphism of $H(D)$ onto the field of complex numbers is evaluation at a point of D. If D is non-empty, prove the kernels of these homomorphisms do *not* exhaust the maximal ideals of the ring $H(D)$.

11. Prove the inversion theorem of Wiener [94]: If f is a continuous function on the unit circle without zeros, and if f has an absolutely convergent Fourier series, then $1/f$ has an absolutely convergent Fourier series. (Hint: Let B be the algebra of all continuous complex-valued functions f on the unit circle for which the Fourier coefficients are absolutely summable:

Analytic Functions with Continuous Boundary Values

$$\sum_{n=-\infty}^{\infty} |c_n| < \infty.$$

Equip B with this sum as norm. Verify that B is a commutative Banach algebra with identity. Show that each homomorphism of B onto the complex numbers is evaluation at a point of the unit circle. Apply the basic result on Banach algebras which we proved.)

12. By the method of Exercise 11, prove the following. If $f(z) = \Sigma\, a_n z^n$ where $\Sigma\, |a_n| < \infty$, and if f has no zeros in the closed unit disc, then $1/f$ has an absolutely convergent Taylor series.

13. Give an example of a non-negative continuous function on the unit circle which has an integrable logarithm but which is not the modulus of a function in A.

CHAPTER 7

THE SHIFT OPERATOR

The Shift Operator on H^2

Many problems in analysis are related to the classification of the invariant subspaces for some bounded linear operator on a Hilbert space. In a 1949 Acta paper, Beurling described all the invariant subspaces for the operator "multiplication by z" on the Hilbert space H^2. This work has been extended in various directions by Lax, by Helson and Lowdenslager, and by Halmos. The work now relates to harmonic analysis on the real line, prediction theory, representations of algebras, representations of semigroups, and the study of function algebras (Dirichlet algebras). One's best guess would be that these extensions of Beurling's work are far from being over. The material we are going to discuss (and the proofs) has now evolved to the point at which one could begin with a brief general discussion of partial isometries on a Hilbert space and then obtain most of the results by specializing the isometries. This is essentially what Halmos has done of late; however, this general point of view does not always give the shortest or most instructive proofs in the special cases. (In some cases it does.) So, we shall begin by discussing Beurling's original problem using the Helson-Lowdenslager proof, and then we shall attempt to describe the various extensions. This will cause some repetition of proofs, but that is probably all for the better.

The linear operator we are going to study is usually called the **shift operator** (on H^2). It is the linear operator T on the Hilbert space H^2, described variously as follows:

(1) H^2 is the space of square-summable sequences of complex numbers:
$$s = [a_0, a_1, a_2, \ldots], \quad \text{and} \quad T(s) = [0, a_0, a_1, \ldots].$$

(2) H^2 is the space of L^2 functions on the unit circle whose Fourier coefficients vanish on the negative integers, and
$$(Tf)(\theta) = e^{i\theta}f(\theta).$$

(3) H^2 is the space of analytic functions in the unit disc for which the

functions $f_r(\theta) = f(re^{i\theta})$ are bounded in L^2 norm, and T is "multiplication by z."

We shall work with the second and third descriptions interchangeably. The problem is this: we wish to find all closed subspaces S of H^2 which are invariant under multiplication by z:

$$zS \subseteq S.$$

An obvious example of such a subspace is the space of functions which vanish at the origin, or the space of functions which vanish at any prescribed set of points in the open disc. Of course, this begins to sound like the ideal theory for the algebra A, which we did in the last chapter. In fact, it sounds so much like it that one is led immediately to a more general example of an invariant subspace:

$$S = FH^2$$

where F is a fixed inner function. But, we would do well to remind ourselves that it was Beurling's inspired observation that inner functions were intimately related to the description of the invariant subspaces for the shift operator. This work was the prelude to his ideal theory for A, not some "aftermath" thereof.

Having observed that each FH^2 is an invariant subspace, one is led to conjecture that there are no other invariant subspaces. This is the case. Assuming that it is so, one must ask: if $S = FH^2$ for some inner function F, how do we find F, given S? We can see that F will be the greatest common divisor of the inner parts of the functions in S. This, too, was Beurling's observation, and it looks natural to us only because we have been through it in the case of continuous boundary values. But he also noted that if $F(0) \neq 0$, then F is a scalar multiple of the orthogonal projection of 1 into S. Suppose F is an inner function and $\lambda = \overline{F(0)}$. Then λF is the orthogonal projection of 1 into FH^2:

$$1 = \lambda F + (1 - \lambda F)$$

$$\frac{1}{2\pi}\int_{-\pi}^{\pi} Fg(1 - \bar{\lambda}\bar{F})d\theta = \frac{1}{2\pi}\int_{-\pi}^{\pi}(F - \bar{\lambda})g d\theta$$

$$= (F(0) - \bar{\lambda})g(0)$$

$$= 0$$

for every g in H^2. This second observation has eventually led to the following proof by Helson and Lowdenslager.

Theorem. *Let S be a non-zero closed subspace of* H^2. *Then S is invariant under multiplication by z if and only if* $S = FH^2$, *where F is an inner function.*

Proof. One half is trivial. Suppose we are given an invariant subspace S. It is no loss of generality to assume that there is at least one function in S which does not vanish at the origin. For, if k denotes the highest order common zero of the functions in S at $z = 0$, then $S = z^k S_0$, where S_0 is invariant and contains a function not vanishing at the origin.

Let G be the orthogonal projection of the constant function 1 into S:

$$1 = G + (1 - G)$$

where G is in S and $(1 - G)$ is orthogonal to S. We are assuming that $G \neq 0$, for 1 is orthogonal to S if and only if every f in S vanishes at the origin. Now we prove that the modulus of G is constant on the unit circle. Since G is in S, we have $z^n G$ in S for $n = 1, 2, 3, \ldots$. By definition, $(1 - G)$ is orthogonal to S, and since $z^n G$ vanishes at the origin, we have

$$0 = \frac{1}{2\pi} \int_{-\pi}^{\pi} (1 - \overline{G}) G e^{in\theta} d\theta$$

$$= \frac{1}{2\pi} \int_{-\pi}^{\pi} G e^{in\theta} d\theta - \frac{1}{2\pi} \int_{-\pi}^{\pi} |G|^2 e^{in\theta} d\theta$$

$$= -\frac{1}{2\pi} \int_{-\pi}^{\pi} e^{in\theta} |G|^2 d\theta.$$

The positive measure $|G|^2 d\theta$ is orthogonal to $e^{in\theta}$, $n = 1, 2, 3, \ldots$ and must, therefore, be a constant multiple of Lebesgue measure.

Now we claim $S = GH^2$. Since the modulus of G is constant on the circle, and since G is in S, it is apparent that S contains GH^2. Let f be a function in S which is orthogonal to GH^2. We prove $f = 0$. Since f is orthogonal to Gz^n, $n = 0, 1, 2, \ldots$, we have

$$0 = \frac{1}{2\pi} \int_{-\pi}^{\pi} f \overline{G} e^{-in\theta} d\theta, \quad n = 0, 1, 2, \ldots.$$

By definition of G we have $(1 - G)$ orthogonal to $z^n f$, i.e.,

$$0 = \frac{1}{2\pi} \int_{-\pi}^{\pi} (1 - \overline{G}) f e^{in\theta} d\theta$$

$$= \frac{1}{2\pi} \int_{-\pi}^{\pi} f e^{in\theta} d\theta - \frac{1}{2\pi} \int_{-\pi}^{\pi} f \overline{G} e^{in\theta} d\theta$$

$$= -\frac{1}{2\pi} \int_{-\pi}^{\pi} f \overline{G} e^{in\theta} d\theta.$$

Therefore, $f\overline{G}$ is the zero function; since G is a non-zero function of constant modulus, this means $f = 0$. We conclude that $S = GH^2$. If we multiply G by a suitable constant, we obtain an inner function F for which $S = FH^2$. Of course, F is unique up to a constant of modulus 1.

Corollary. *If* S *is a non-trivial closed subspace of* H^2 *which is invariant under multiplication by* z*, then* S = FH^2*, where* F *is the greatest common divisor of the inner parts of the functions in* S.

Corollary. *Let* f *be a function in* H^2. *Then the functions* $z^n f$, $n = 0, 1, 2, \ldots$, *span* H^2 *if and only if* f *is an outer function, that is, if and only if* $f \neq 0$ *and*

$$\log |f(0)| = \frac{1}{2\pi} \int_{-\pi}^{\pi} \log |f(e^{i\theta})| d\theta.$$

Proof. The span of the functions $z^n f$ is simply the smallest closed subspace of H^2 which contains f and is invariant under multiplication by z. This subspace is H^2 if and only if the inner parts of the functions in it have greatest common divisor 1, and this clearly means that the inner part of f is 1, i.e., that f is an outer function. When we first introduced outer functions, we observed that they are characterized by Jensen's inequality being equality.

We should probably comment that the characterization of these invariant subspaces exhibits explicitly their lattice structure. To each non-zero invariant subspace S we have assigned a unique inner function F. If another such subspace S_1 is given, it is clear that S is contained in S_1 if and only if F_1 divides F. Every collection of these invariant subspaces has a least upper bound, namely, the subspace corresponding to the greatest common divisor of the associated inner functions. The process is, of course, reversible. That is, one can use this description of invariant subspaces to prove that any collection of inner functions has a greatest common divisor.

More about Dirichlet Algebras

The central part of the characterization above extends to the context of Dirichlet algebras, which we introduced in Chapter 4. Recall that a Dirichlet algebra is a uniformly closed algebra A of continuous complex-valued functions on a compact Hausdorff space X, such that the real parts of the functions in A are dense in the real continuous functions on X. If we have any non-zero positive measure m on X which is *multiplicative* on A, we define $H^2(dm)$ to be the closure in $L^2(dm)$ of the functions in A. Here we want to comment on "invariant subspaces" of $H^2(dm)$. Of course, we do not expect a shift operator on $H^2(dm)$. The subspaces we discuss are those which are invariant under multiplication by every function f in A.

If A is the algebra of continuous functions on the unit circle whose Fourier coefficients vanish on the negative integers, and if m is normalized Lebesgue measure, then $H^2(dm) = H^2$, and the subspaces invariant under multiplication by functions in A are just the invariant subspaces for the shift operator. Thus, the situation we describe is a generalization of the study of invariant subspaces for the shift operator.

Suppose we are given the Dirichlet algebra A on the compact space X and we fix a positive measure m on X which is multiplicative on A.

Theorem. *Let S be a closed subspace of $H^2(dm)$ which is invariant under multiplication by the functions in A. Suppose that there is at least one function g in S such that $\int g dm \neq 0$. Then there exists a function F in $H^2(dm)$ which has modulus 1 almost everywhere with respect to m and for which $S = FH^2(dm)$.*

Proof. We sketch the proof, which is virtually identical to the proof we gave for the corresponding theorem in the disc. First, one can easily verify that m is multiplicative on $H^2(dm)$. If the subspace S is given as above, our hypothesis says that 1 is not orthogonal to S. Let G be the orthogonal projection of 1 into S. Let A_0 denote the set of functions f in A for which $\int f dm = 0$. Since

$$\int Gf dm = \int f dm \cdot \int G dm = 0$$

for f in A_0, we have

$$0 = \int (1 - \bar{G})Gf dm$$
$$= -\int f|G|^2 dm$$

for every f in A_0. Since A is a Dirichlet algebra, the measure $|G|^2 dm$ is a constant multiple of dm. So $|G| = k$, where k is a non-zero constant. It is then clear that S contains $GH^2(dm)$. If g is in S and is orthogonal to $GH^2(dm)$ we have

$$\int g\bar{G}\bar{f} dm = 0, \quad f \text{ in } A.$$

But by the definition of G, we have

$$0 = \int (1 - \bar{G})fg dm$$
$$= -\int g\bar{G}f dm$$

if f is in A_0. Thus, the measure $g\bar{G}dm$ is zero, which says that $g = 0$ almost everywhere dm.

Those subspaces of $H^2(dm)$ in which all the functions vanish at m cannot be taken care of by a function of modulus 1. For example, let X be the torus. Choose an irrational number α, and let A be the algebra of all continuous functions on the torus whose Fourier coefficients

$$a_{mn} = \frac{1}{4\pi^2} \int_{-\pi}^{\pi} \int_{-\pi}^{\pi} f(\theta, \psi) e^{-im\theta} e^{-in\psi} d\theta d\psi$$

vanish outside the half-plane where $m + n\alpha \geq 0$. Then A is a Dirichlet algebra. If $dm = \frac{1}{4\pi^2} d\theta d\psi$, $H^2(dm)$ becomes the space of square-summable functions on the torus with a Fourier series

$$f(\theta, \psi) \sim \sum_{m + n\alpha \geq 0} a_{mn} e^{im\theta} e^{in\psi}.$$

The Shift Operator

If we take S to be the subspace of functions f for which $a_{00} = 0$, then S is invariant under multiplication by functions in A, but is not of the form $S = FH^2(dm)$.

The criterion for a function to generate $H^2(dm)$ also generalizes, as follows.

Theorem. *Let g be a function in $H^2(dm)$. Then the closed linear span of the functions fg with f in A is $H^2(dm)$ if and only if*

$$\int \log |g| dm = \log |\int g \, dm| > -\infty.$$

Proof. The proof can be given rather easily with the use of Szegö's theorem for Dirichlet algebras, which we proved in Chapter 4. Let g be a function in $H^2(dm)$. Suppose that Ag is dense in $H^2(dm)$. Then there is a sequence of functions g_n in A such that

$$\int |1 - g_n g|^2 dm \to 0.$$

Certainly then,

$$\int g \, dm = \lambda \neq 0 \quad \text{and} \quad \int g_n \, dm \to \lambda^{-1}.$$

We can, therefore, assume that $g_n = \lambda^{-1} - f_n$, where f_n is in A_0, the set of f in A such that $\int f \, dm = 0$. Now

$$\int |1 - (\lambda^{-1} - f_n)g|^2 dm = -1 + \int |\lambda^{-1} - f_n|^2 |g|^2 dm$$
$$= -1 + |\lambda|^{-2} \int |1 - \lambda f_n|^2 |g|^2 dm.$$

We conclude that

$$\int |1 - \lambda f_n|^2 |g|^2 dm \to |\lambda|^2.$$

Thus

$$\inf_{f \in A_0} \int |1 - f|^2 |g|^2 dm \leq |\lambda|^2 = |\int g \, dm|^2.$$

But the reverse inequality holds for any g. Thus the infimum is equal to $|\lambda|^2$. Szegö's theorem states that this infimum is

$$\exp \left[\int \log |g|^2 dm\right]$$

so we obtain

$$\int \log |g| dm = \log |\int g \, dm| > -\infty.$$

It is very easy to reverse the steps if this last condition holds, to conclude that 1 is in the closure of Ag; hence, Ag spans $H^2(dm)$.

Invariant Subspaces for H^2 of the Half-plane

Consider the half-plane $Re(w) > 0$. If f is analytic in this right half-plane, we say that f is in the class H^2, provided that the integrals

$$\int_{-\infty}^{\infty} |f(x + iy)|^2 dy$$

are bounded for $x > 0$. A theorem of Paley and Wiener states that each such f has non-tangential limits at almost every point on the imaginary axis, that these boundary values are square-integrable, and that f is the Poisson integral of its boundary values:

$$f(x + iy) = \frac{1}{\pi} \int_{-\infty}^{\infty} f(it) \frac{x}{x^2 + (y - t)^2} dt.$$

Let us assume this theorem for the moment. (We shall prove it in the next chapter.) Then H^2 of the half-plane becomes a Hilbert space, with the inner product

$$(f, g) = \int_{-\infty}^{\infty} f(it)\overline{g(it)}dt.$$

In a 1959 Acta paper, Peter Lax extended Beurling's result about the invariant subspaces for the shift operator to certain "invariant" subspaces of H^2 of the half-plane. Actually, Lax considered vector-valued analytic functions, which we shall discuss later. We wish now to discuss the problem for the scalar-valued H^2 described above, and to show that this result of Lax is equivalent to Beurling's result.

Lax's scalar-valued theorem is the following. If S is a closed subspace of H^2 of the right half-plane, then S is invariant under multiplication by the functions $e^{-\lambda w}$, $\lambda \geq 0$, if and only if $S = FH^2$, where F is an inner function (i.e., F is analytic in the half-plane, bounded by 1, and has non-tangential limits which exist and are of modulus 1 almost everywhere on the imaginary axis). If one uses a stronger form of the Paley-Wiener theorem (which we shall not), every function f in H^2 has the form

$$f(w) = \int_0^\infty \tilde{f}(x) e^{-xw} dx$$

where $\tilde{f}$ is square-integrable; that is, every f in H^2 is the Laplace (or Fourier) transform of a square-integrable function on the positive half-line. A subspace S of H^2 which is invariant under multiplication by $e^{-\lambda w}$ for all $\lambda \geq 0$ is then simply the "transform" of a subspace of $L^2(0, \infty)$ which is invariant under all right translations.

We want to establish the simple relationship between H^2 of the disc and H^2 of the half-plane. If we map the disc onto the right half-plane by

$$w = \frac{1 + z}{1 - z}$$

then H^2 of the disc is carried onto a space of analytic functions in the half-plane; we wish to relate that space to H^2 of the half-plane. On the boundary the linear fractional map is

$$it = \frac{1 + e^{i\theta}}{1 - e^{i\theta}}$$

from which it is easy to obtain

$$\frac{1}{2\pi} d\theta = \frac{1}{\pi}(1 + t^2)^{-1}dt.$$

Thus, if g is measurable on the circle and $f(it) = g(e^{i\theta})$, we see that g is integrable if and only if f is integrable with respect to the measure $(1+t^2)^{-1}dt$; when g is integrable,

$$\frac{1}{2\pi}\int_{-\pi}^{\pi} g(e^{i\theta})d\theta = \frac{1}{\pi}\int_{-\infty}^{\infty} f(it) \cdot \frac{1}{1+t^2} dt.$$

In particular, this will hold for g in H^2. When g is in H^2, its values inside the disc are given by

$$g(z) = \frac{1}{2\pi}\int_{-\pi}^{\pi} g(e^{i\theta}) \operatorname{Re}\left[\frac{e^{i\theta}+z}{e^{i\theta}-z}\right] d\theta.$$

If we define $f(w) = g(z) = g\left(\dfrac{w-1}{w+1}\right)$, then this Poisson integral formula for g transforms into the Poisson formula

$$f(w) = \frac{1}{\pi}\int_{-\infty}^{\infty} f(it) \operatorname{Re}\left[\frac{itw-1}{it-w}\right] \cdot \frac{1}{1+t^2} dt$$

or

$$f(x+iy) = \frac{1}{\pi}\int_{-\infty}^{\infty} f(it) \frac{x}{x^2+(y-t)^2} dt.$$

Thus, H^2 of the disc is transformed into the space of analytic functions in the right half-plane which are the Poisson integral (as above) of a function on the imaginary axis which is in $L^2\left(\dfrac{1}{1+t^2} dt\right)$. In other words, as a subspace of L^2 of the circle, H^2 is transformed into the subspace of $L^2\left(\dfrac{1}{1+t^2} dt\right)$ of those functions whose Poisson integrals are analytic for $\operatorname{Re} w > 0$.

Now H^2 of the right half-plane consists of all functions analytic for $\operatorname{Re} w > 0$ which are the Poisson integral of a function on the imaginary axis which is in $L^2(dt)$. Certainly, then, if f is in H^2 of the half-plane, the function

$$g(z) = f\left(\frac{1+z}{1-z}\right)$$

is in H^2 of the disc. But we claim that $\dfrac{1}{1-z} g(z)$ is in H^2 of the disc. For

$$\frac{1}{1-z} g(z) = \tfrac{1}{2}(1+w)f(w) \qquad \left[w = \frac{1+z}{1-z}\right].$$

Since $f(it) \in L^2(dt)$ and $|1+it|^2 = 1+t^2$, we see that $(1+it)f(it)$ is in $L^2\left(\dfrac{1}{1+t^2} dt\right)$. Thus, $\dfrac{1}{1-e^{i\theta}} g(e^{i\theta})$ is square-integrable on the circle. It is a simple matter to verify that if g is in H^2 of the disc and $\dfrac{1}{1-e^{i\theta}} g(e^{i\theta})$

is in L^2 of the circle, then $\dfrac{1}{1-z}g(z)$ is in H^2. We conclude that, if f is in H^2 of the right half-plane, then $(1+w)f(w)$ is, under the map $w = \dfrac{1+z}{1-z}$, the image of an H^2 function in the unit disc. The converse of this also holds. Suppose we have f analytic for Re $w > 0$ and $(1+w)f(w)$ comes from an H^2 function in the disc. This means only that the boundary values $(1+it)f(it)$ exist almost everywhere, that they are in $L^2\left(\dfrac{1}{1+t^2}dt\right)$, and that $(1+w)f$ is the Poisson integral of these boundary values. Certainly, then, $f(it) \in L^2(dt)$. Also, f is the Poisson integral of its boundary values, because f obviously comes from an H^2 function in the disc if $(1+w)f(w)$ does (division by $(1+w)$ is essentially multiplication by $(1-z)$ in the disc). What we have proved, assuming the Paley-Wiener theorem, may be summarized as follows:

Theorem. *Let* f *be an analytic function in the right half-plane. Then* f *is in* H^2 *if and only if the function* $(1+w)f(w)$ *is transformed by the map* $z = \dfrac{w-1}{w+1}$ *into a function* h *in* H^2 *of the unit disc. Indeed, the map from* f *to* $\sqrt{\pi}\,h$ *is an isometry of* H^2 *onto* H^2 *of the disc.*

In order to handle "invariant" subspaces we need two simple lemmas.

Lemma. *Let* S *be a closed subspace of* H^2 *of the disc. Then* S *is invariant under multiplication by* z *if and only if* S *is invariant under multiplication by every bounded analytic function.*

Proof. Of course, this is obvious from Beurling's characterization of the invariant subspaces for multiplication by z; however, it is also obvious a priori. One half is trivial. So, suppose S is invariant under multiplication by z. Clearly, S is invariant under multiplication by any polynomial in z; hence, S is invariant under multiplication by f, where f is any uniform limit of polynomials. Now given any $f \in H^\infty$, if $r < 1$ the function $f_r(z) = f(rz)$ is a uniform limit of polynomials. If $g \in S$, we have $f_r g$ in S for each $r < 1$. If h is in $L^2(d\theta)$, and if

$$\int f_r g h d\theta = 0$$

for each $r < 1$, then $\int fgh d\theta = 0$, since $f_r \to f$ boundedly and pointwise almost everywhere. Thus, fg is in the L^2 closed span of $\{f_r g\}$. Consequently, S is invariant under multiplication by H^∞. (One can replace f_r by σ_n, the nth Cesaro mean for f, and the argument goes as well.)

Lemma. *Let* S *be a closed subspace of* H^2 *of the right half-plane. Then* S *is invariant under multiplication by the functions* $e^{-\lambda w}$, $\lambda \geq 0$, *if and only if*

S *is invariant under multiplication by every bounded analytic function in the half-plane.*

Proof. Again, one half is trivial. Suppose S is invariant under multiplication by $e^{-\lambda w}$ for every $\lambda \geqq 0$. Let f be a bounded analytic function. We wish to show that S is invariant under multiplication by f. From what we did above in the disc, we know that f is the bounded pointwise (a.e.) limit of a sequence of polynomials in $\dfrac{w-1}{w+1}$. Thus, if we prove the invariance for $f(w) = \dfrac{w-1}{w+1}$, we shall be done. Actually, $f(w) - 1 = \dfrac{-2}{w+1}$ will do as well. We see that we need only prove that S is invariant under multiplication by $(1+w)^{-1}$. Now

$$\frac{1}{1+w} = \int_0^\infty e^{-(1+w)x} dx.$$

Let
$$f_n(w) = \int_0^n e^{-(1+w)x} dx.$$

Approximating sums for the integral show that f_n is a bounded pointwise limit of linear combinations of the $e^{-\lambda w}$, $\lambda \geqq 0$. Thus S is invariant under multiplication of f_n. But $f_n(w) \to (1+w)^{-1}$ pointwise and the convergence is bounded. That proves the lemma.

Theorem (Lax). *Let* S *be a non-zero closed subspace of* H^2 *of the right half-plane, and suppose that* S *is invariant under multiplication by* $e^{-\lambda w}$ *for every* $\lambda \geqq 0$. *Then* S *has the form* S = FH^2, *where* F *is an inner function.*

Proof. We look at the Hilbert space

$$\tilde{H} = (1+w)H^2$$

with the inner product

$$(f, g) = \frac{1}{\pi} \int_{-\infty}^\infty f(it)\overline{g(it)} \frac{1}{1+t^2} dt.$$

We proved above that $f(w) \to \sqrt{\pi}(1+w)f(w)$ is an isometry of H^2 onto $\tilde{H}$, and that $\tilde{H}$ is precisely the image of H^2 of the disc under the linear fractional map $z = \dfrac{w-1}{w+1}$. Obviously, $\tilde{S} = (1+w)S$ is a closed subspace of $\tilde{H}$.

From the last lemma, we know that S is invariant under multiplication by any bounded analytic function. Clearly, $\tilde{S}$ has the same property. Under the linear fractional map, $\tilde{H}$ goes onto H^2 of the disc and $\tilde{S}$ onto a closed subspace thereof. Since this map preserves the class of bounded analytic functions, we have immediately from Beurling's result that $\tilde{S} = F\tilde{H}$, where F is an inner function. Thus, $S = FH^2$.

Several remarks are in order. First, as we commented earlier, the discussion above shows the Beurling and Lax theorems to be equivalent. Second, the inner function associated with S is unique up to a constant of modulus 1. Third, one of the ideas which helps to unify the theorems in the disc and half-plane is that of considering subspaces invariant under multiplication by all bounded analytic functions. These theorems seem to deal with special types of representations of the algebra of bounded analytic functions, which we shall discuss later.

Isometries

Recently, Halmos has observed that one or two elementary results about isometries on a Hilbert space shed considerable light on the type of invariant subspace problem which we have been discussing. These results do not simplify the characterization of the invariant subspaces for the shift operator; they do, however, lend perspective to the discussion. Also, they greatly simplify the proof of Lax's vector-valued theorem, and they facilitate the discussion of the invariant subspaces for multiplication by z on L^2 of the circle. In the latter context, the ideas are very similar to those used by Helson and Lowdenslager.

Let H be a Hilbert space. An **isometry** on H is a linear transformation (operator) from H into H which preserves inner products:

$$(Tx, Ty) = (x, y).$$

If T maps H onto H, we call T a **unitary** operator. The canonical example of an isometry which is not unitary is the shift operator on H^2, or multiplication by z on H^2 of the disc. A slightly more general example is the following. Let K be an arbitrary Hilbert space. Let $H^2(K)$ be the space of sequences

$$h = [h_0, h_1, h_2, \ldots]$$

of elements of K for which

$$\sum_{n=0}^{\infty} ||h_n||^2 < \infty.$$

The inner product on $H^2(K)$ is

$$(g, h) = \sum_{n=0}^{\infty} (g_n, h_n).$$

Let T be the "shift" operator on $H^2(K)$:

$$T(h_0, h_1, \ldots) = (0, h_0, h_1, \ldots).$$

It is clear that T is a non-unitary isometry on $H^2(K)$. Of course, the shift operator on H^2 is the special case of this example when $K = C$, the complex numbers.

The Shift Operator

Lemma. *Let* T *be an isometry on a Hilbert space* H, *and let* N *be the orthogonal complement of the range of* T. *Then the subspaces*
$$T^k(N), \quad k = 0, 1, 2, 3, \ldots$$
are pairwise orthogonal.

Proof. We should note that since T is an isometry the subspaces N, $T(N)$, $T^2(N)$, ... are closed. For the proof, suppose $0 \leq j < k$. Let $x, y \in N$. We wish to show that $(T^j x, T^k y) = 0$. Since T is isometric and $k > j$,
$$(T^j x, T^k y) = (x, T^{(k-j)} y)$$
and the latter inner product is zero because $T^{(k-j)}y$ is in the range of T and x is in N.

Theorem. *Let* T *be an isometry on a Hilbert space* H. *Let* N *be the orthogonal complement of the range of* T, *and let* M *be the orthogonal complement of the span of the spaces* $T^k(N)$, k $\geq$ 0. *Then*

(i) H = M $\oplus$ N $\oplus$ T(N) $\oplus$ T²(N) $\oplus$ $\cdots$.

(ii) *the subspace* M *is invariant under* T; *indeed, the restriction of* T *to* M *is a unitary operator on* M.

(iii) *the subspace* M *consists of all vectors* h *in* H *which are "infinitely divisible" by* T, *i.e., all* h *such that* h *is in the range of* T^k *for every nonnegative integer* k.

Proof. By the lemma above, the spaces $T^k(N)$, $k \geq 0$, are pairwise orthogonal. By the definition of M we then have (i). Suppose $m \in M$. Since m is orthogonal to N, we must have m in the range of T, i.e., $m = Th$ for some h in H. But h is also in M; for if $k \geq 0$ and $n \in N$,
$$(T^k n, h) = (T^{k+1} n, Th)$$
$$= (T^{k+1} n, m)$$
$$= 0.$$

Thus $M \subseteq T(M)$. Note that this shows that every $m \in M$ is "infinitely divisible" by T. It is also easy to see that M is invariant under T. If $m \in M$ and $n \in N$, the inner product
$$(T^k n, Tm)$$
is zero for $k \geq 1$ because $m \in M$, and is zero for $k = 0$ by definition of N. That proves (ii). We have already proved one half of (iii), that the vectors in M are infinitely divisible by T. Conversely, if $h \in H$, and if for each $k \geq 0$ there is an h_k in H with $h = T^k(h_k)$, it is very easy to check that $h \in M$. For example, if $h = Th_1$, we have h orthogonal to N. If $h = T^2 h_2$, then h is orthogonal to N and $T(N)$, etc.

Corollary. *Let* T *be an isometry on a Hilbert space* H. *Then there are subspaces* M *and* N *of* H *such that* T *is the direct sum of a unitary operator* on M *and an operator on* $M^\perp$ *which is unitarily equivalent to the shift operator on* $H^2(N)$.

Proof. Let M and N be as above. Let T_1 be the restriction of T to M and T_2 the restriction of T to $M^\perp$. Then T_1 is unitary and T is the direct sum of T_1 and T_2. Now
$$M^\perp = N \oplus T(N) \oplus T^2(N) \oplus \cdots.$$
There is a completely obvious isomorphism of $M^\perp$ with $H^2(N)$ which carries T_2 onto the shift operator on $H^2(N)$.

Corollary. *Let* T *be an isometry on a Hilbert space* H. *The following three properties of* T *are equivalent:*

(i) *There is no non-zero subspace of* H *on which* T *is unitary.*

(ii) *There is no non-zero element of* H *which is "infinitely divisible" by* T.

(iii) *There is a Hilbert space* N *such that* T *is unitarily equivalent to the shift operator on* $H^2(N)$.

Proof. We have done the more difficult part of the work above. That is, we have shown that (i) and (ii) are equivalent and that each implies (iii). The argument will be completed if we show that the shift operator on $H^2(N)$ has property (ii). This is obvious.

Now we have characterized the shift operator on H^2 of some Hilbert space by the properties of being (a) an isometry; (b) unitary on no non-zero subspace. This makes it very easy to give (à la Halmos) the description of the invariant subspaces for this operator, which were first found by Lax. First, we shall show how we can abstractly characterize the shift operator on H^2 (of the complex numbers). When we have done this, we turn (in the next section) to the application of these ideas to the characterization of all invariant subspaces for the shift operator on L^2 of the circle. Then we return to the invariant subspaces for the shift on H^2 of a Hilbert space.

Theorem. *Let* T *be a non-unitary isometry on a Hilbert space* H. *The following are equivalent:*

(i) T *is unitarily equivalent to the shift operator on* H^2 (*i.e., equivalent to multiplication by* z *on* H^2 *of the disc*).

(ii) *There is no non-trivial subspace of* H *which completely reduces* T (*i.e., if* S *is a closed subspace of* H *such that* S *and* $S^\perp$ *are invariant under* T, *then* $S = \{0\}$ *or* $S = H$).

Proof. Let us first show that multiplication by z on H^2 has property (ii). It will suffice to show that if $f, g \in H^2$, and if $z^m f$ is orthogonal to $z^n g$ for all $m, n \geqq 0$, then either $f = 0$ or $g = 0$. But this is clear, for we have
$$\int_{-\pi}^{\pi} e^{i(m-n)\theta} f(\theta)\overline{g(\theta)} d\theta = 0, \quad m, n = 0, 1, 2, \ldots.$$

Thus, $f\bar{g} = 0$ almost everywhere on the circle. So either f or g must vanish on a set of positive measure on the circle, and for an H^2 function this makes it identically zero.

Now suppose we have a non-unitary isometry T on H with no (non-trivial) completely reducing subspace. In particular, if we let N be the orthogonal complement of the range of T, we must have

$$H = N \oplus T(N) \oplus T^2(N) \oplus \cdots.$$

We cannot have $N = 0$, since T is not unitary; hence, in the decomposition

$$H = M \oplus N \oplus T(N) \oplus \cdots$$

we must have $M = \{0\}$. In other words, (ii) guarantees that T "is" the shift operator on $H^2(N)$, and our task is now to prove that N is one-dimensional. Let n be a non-zero vector in N, and let S be the T-invariant subspace spanned by n, i.e., the span of the vectors $T^k n$, $k \geq 0$. We claim that $S^\perp$ is invariant under T. Now $h \in S^\perp$ means that

$$(T^k n, h) = 0 \quad \text{for } k = 0, 1, 2, \ldots.$$

Thus, for $k = 1, 2, 3, \ldots$ we have $(T^k n, Th) = 0$. But $(n, Th) = 0$ by definition of N, so Th is in $S^\perp$. Since $S \neq \{0\}$, we must have $S^\perp = \{0\}$ by (ii). Thus, $S = H$; i.e., every h in H is uniquely expressible in the form

$$h = a_0 n + a_1 T n + a_2 T^2 n + \cdots$$

where $\Sigma |a_k|^2 < \infty$. In particular, it is clear that every vector in N is a scalar multiple of n.

The Shift Operator on L^2.

Now we turn to the shift operator (multiplication by $e^{i\theta}$) on L^2 of the unit circle. We shall describe the invariant subspaces for this operator. Of course, one invariant subspace is H^2, or, more generally, any subspace of H^2 which is invariant under multiplication by z. Thus the discussion is a slight extension of the characterization of the invariant subspaces for the shift operator on H^2.

Let S be a closed subspace of L^2 of the circle such that $e^{i\theta} S \subseteq S$. It may happen that S is also invariant under multiplication by $e^{-i\theta}$. This means only that S is a subspace on which the shift operator is unitary, or that S is invariant under multiplication by every bounded measurable function. It is well-known that any such subspace has the form $S = \chi L^2$ where χ is the characteristic function of some Baire set; in other words, such an S consists of all functions in L^2 which vanish on some fixed Baire set.

Theorem. *Let* S *be a closed subspace of* L^2 *which is invariant under multiplication by* z.

(i) *If $zS = S$, then S consists of all functions in L^2 which vanish on some fixed Baire set on the circle.*

(ii) *If $zS \neq S$, then $S = FH^2$, where F is a measurable function of modulus 1.*

Proof. As we said, (i) is well-known and easy to verify. Suppose $zS \neq S$. Let N be the orthogonal complement of zS in S. Then

$$S = M \oplus N \oplus zN \oplus z^2N \oplus \cdots$$

where M consists of all functions in S which are "infinitely divisible" by z in S. Suppose f is in N. Then f is orthogonal to $z^k f$ for $k = 1, 2, 3, \ldots$ or

$$\int_{-\pi}^{\pi} |f(\theta)|^2 e^{ik\theta} d\theta = 0, \quad k = 1, 2, 3, \ldots.$$

Thus $|f|$ is constant almost everywhere on the circle.

Now M is a subspace of L^2 on which multiplication by z is unitary. Thus $M = \chi L^2$, where χ is the characteristic function of some Baire set. Choose $f \neq 0$ in N. There is such an f since $zS \neq S$. Then $z^k f$ is orthogonal to χ for $k = 0, 1, 2, \ldots$; i.e.,

$$\int_{-\pi}^{\pi} e^{ik\theta} f(\theta) \chi(\theta) d\theta = 0, \quad k = 0, 1, 2, \ldots.$$

Thus $f\chi$ is in H^2. Obviously, $\chi \neq 1$, so $f\chi$ vanishes on a set of positive measure. But $f\chi$ is in H^2, so we must have $f\chi = 0$ a.e. On the other hand, f is a non-zero function of constant modulus. We conclude that $\chi = 0$.

Now we have

$$S = N \oplus zN \oplus z^2N \oplus \cdots.$$

The subspace N is one-dimensional. There are several ways to see this. First, it is evident from the fact that each function in N has constant modulus. Second, if we have functions f and g in N orthogonal to one another, we have f orthogonal to $z^k g$ for $k \geq 0$ and $z^k f$ orthogonal to g for $k \geq 0$ so that $f\bar{g} = 0$ and (by the constant modulus property) either $f = 0$ or $g = 0$. Third, by the general result of the last section, it would suffice to show that there is no non-trivial subspace of S which completely reduces multiplication by z. This means that if $f, g \in S$ and if $z^j f$ is orthogonal to $z^k g$ for all $j, k \geq 0$, then either $f = 0$ or $g = 0$; that is, if $f, g \in S$ and $f\bar{g} = 0$, then $f = 0$ or $g = 0$. This can be done by showing at the outset that any function in S which vanishes on a set of positive measure is identically zero.

Since N is one-dimensional, we can choose an F in N of modulus 1, and S will consist of all functions of the form

$$a_0 F + a_1 z F + a_2 z^2 F + \cdots$$

where $\Sigma |a_n|^2 < \infty$, i.e., $S = FH^2$.

There is an analogous theorem for L^2 of the real line. If we use the Paley-Wiener theorem and the relation between H^2 of the disc and H^2 of the half-plane, we obtain this result immediately. With each g in L^2 of the circle we associate a function f on the imaginary axis by

$$f(it) = \frac{1}{1+it} g\left(\frac{it-1}{it+1}\right).$$

Then f is in $L^2(dt)$. In fact, up to a constant, $g \to f$ is an isometry of L^2 of the circle onto L^2 of the line. This map carries H^2 of the circle onto H^2 of the imaginary axis, i.e., the boundary values of the functions in H^2 of the right half-plane. In the last theorem we were studying the subspaces of L^2 of the circle invariant under multiplication by z, i.e., invariant under multiplication by (the boundary values of) every bounded analytic function. These subspaces are carried onto the subspaces of $L^2(dt)$ invariant under multiplication by such boundary values, or onto the subspaces of $L^2(dt)$ invariant under multiplication by $e^{-i\lambda t}$ for all $\lambda \geqq 0$. The result from the disc now carries over directly. It perhaps sounds more natural if we perform a 90-degree rotation and use the upper half-plane.

Theorem. *Let* H^2 *denote the space of square-integrable functions on the real line which are boundary values of functions in* H^2 *of the upper half-plane. Let* S *be any closed subspace of* L^2 *of the line which is invariant under multiplication by* $e^{i\lambda x}$ *for all* $\lambda \geqq 0$. *Then* S *is of one of the following two types:*
  (i) S *consists of all functions in* L^2 *which vanish on some fixed Baire set.*
  (ii) S $=$ FH^2, *where* F *is a measurable function of modulus* 1.

If one uses the full strength of the theorem of Paley and Wiener which we have been discussing, there results a characterization of all subspaces of L^2 of the line which are invariant under right translation. This amounts to the use of the Plancherel theorem. If f is an integrable function on the line, the **Fourier transform** of f is the function $\hat{f}$ defined by

$$\hat{f}(x) = \frac{1}{\sqrt{2\pi}} \int_{-\infty}^{\infty} f(t)e^{ixt}dt.$$

The Plancherel theorem states that (a) if f is in $L^1 \cap L^2$ then $\hat{f}$ is in L^2 and $||\hat{f}||_2 = ||f||_2$, using the measure $\frac{1}{\sqrt{2\pi}}dt$; (b) the map from $L^1 \cap L^2$ into L^2 defined by $f \to \hat{f}$ has a unique extension to a unitary map of L^2 onto L^2. This extension defines the Fourier transform of any L^2 function. The Paley-Wiener theorem says that the space H^2 of the last theorem is the set of all Fourier transforms of L^2 functions which vanish on the half-line $(-\infty, 0)$. Translation of f by λ multiplies the Fourier transform by $e^{i\lambda x}$. The last theorem may then be stated as follows.

Theorem. *Let H^2 denote the space of Fourier transforms of L^2 functions which vanish on the left half-line. Suppose S is a (closed) subspace of L^2 which is right translation invariant.*

(i) If S is invariant under all translations, then S consists of all functions in L^2 whose Fourier transforms vanish on some fixed Baire set E', i.e., $\hat{S} = \chi_E L^2$.

(ii) Otherwise, $\hat{S} = FH^2$, where F is a measurable function of modulus 1.

We might mention one interesting corollary to the description of these invariant subspaces of L^2. Suppose f is an L^2 function on the unit circle. Then the functions $e^{in\theta}f(\theta)$ with $n \geq 0$ span L^2 if and only if

(a) f does not vanish on a set of positive measure;

(b) $\log |f|$ is *not* integrable.

Equivalently, suppose f is in L^2 of the line. Then the right translates of f span L^2 if and only if

(a) the Fourier transform $\hat{f}$ does not vanish on a set of positive measure;

(b) $\dfrac{\log |\hat{f}(x)|}{1+x^2}$ is *not* Lebesgue integrable.

Actually, these results were known prior to the characterization of the "invariant" subspaces. They are corollaries to Szegö's theorem, which we proved in Chapter 4.

The Vector-valued Case

We shall now give the description of the invariant subspaces for the shift operator on H^2 of a Hilbert space. This was first done by Lax. We shall follow Halmos in obtaining the result as an easy corollary to the results on isometries which we have already given. The discussion here will be brief. The proof, as such, is complete; however, we shall skim over some standard preliminaries to avoid becoming embroiled in a lengthy discussion of vector-valued integration and analytic functions with values in a Banach space.

Recall that for any Hilbert space K we defined $H^2(K)$ as the space of all sequences
$$h = [h_0, h_1, h_2, \ldots]$$
of elements of K for which $\Sigma \, ||h_n||^2 < \infty$. Each element of $H^2(K)$ may also be interpreted as an analytic function in the unit disc with values in K:
$$h(z) = \sum_{n=0}^{\infty} z^n h_n.$$

This amounts to regarding h as a square-integrable function on the circle with values in K, and then extending h to the disc by the Poisson integral formula. The functions we obtain in the disc are characterized by the property that the integrals

$$\int_{-\pi}^{\pi} ||h(re^{i\theta})||^2 d\theta$$

are bounded as $r \to 1$. Each such function in the disc may be identified with its boundary values as in the scalar-valued case, and we may thus identify $H^2(K)$ with this space of analytic functions in the unit disc. Naturally, this identifies the shift operator with "multiplication by z." The "invariant subspace theorem" then assumes the following form.

Theorem. *Let K be a Hilbert space, and let S be a closed non-zero subspace of $H^2(K)$ which is invariant under multiplication by z. Then there exists a Hilbert space N and a function F such that*

(i) *F is an analytic function in the unit disc with values in the space of bounded linear operators from N into K; if $|z| < 1$ then $||F(z)|| \leq 1$; at almost every point $e^{i\theta}$ on the unit circle $F(e^{i\theta})$ is an isometry;*

(ii) $S = FH^2(N)$; *i.e., S consists of all g in $H^2(K)$ of the form*

$$g(z) = F(z)f(z)$$

where f is in $H^2(N)$.

Proof. If we restrict "multiplication by z" to the subspace S, we obtain an isometry on S which is not unitary on any subspace of S. Thus, if we let N denote the orthogonal complement in S of the subspace zS, we shall have

$$S = N \oplus zN \oplus z^2 N \oplus \cdots.$$

This enables us to identify S with $H^2(N)$. The function F is defined as follows. Suppose $|z| < 1$. Then $F(z)$ is the linear operator from N into K obtained by evaluating each element of N at the point z. Now $S = FH^2(N)$ because of the above direct sum decomposition for S. The other properties of F stated in (i) are easy to verify, modulo the preliminaries we said we would skim over. Any two reasonable-sounding definitions of an analytic function with values in a Banach space are equivalent. Here one can use the usual existence of the derivative. Each $h \in H^2(K)$ is differentiable because it is the sum of a convergent power series. It is, then, easy to see that F is differentiable, $F'(z)$ being the operator which sends each n in N into $n'(z)$. Obviously,

$$||F(z)|| \leq 1 \quad \text{for } |z| < 1$$

so F has boundary values at almost every point of the circle. This is the analogue of Fatou's theorem on bounded scalar-valued analytic functions, and its proof can be given in the same manner. To check that these boundary values are isometric almost everywhere, just use the fact that each n in N is of constant norm almost everywhere on the circle (n is orthogonal to $z^k n$ for $k = 1, 2, 3, \ldots$).

We should remark that the subspace N has dimension not greater than the dimension of the underlying Hilbert space K. The theorem would seem more elegant if N were always isomorphic to K; however, the simplest examples show that $\dim N < \dim K$ occurs. The result will seem to be a more natural generalization of the scalar-valued case if we "embed" N in K. Choose an isometry of N onto some closed subspace K_0 of K. Then regard the function F as having values in the bounded operators from K_0 into K. For each z define $F(z)$ to be 0 and $K_0^\perp$, thereby extending $F(z)$ to a bounded linear operator on K. On the boundary, the operator $F(e^{i\theta})$ is (almost everywhere) a partial isometry on K, that is, an operator which is isometric on a subspace and zero on its orthogonal complement. The theorem then states that every subspace of $H^2 = H^2(K)$ which is invariant under multiplication by z has the form FH^2, where F is an analytic function in the unit disc whose values are bounded operators on K; $\|F(z)\| \leq 1$; and at almost every point of the unit circle $F(e^{i\theta})$ is a partial isometry.

It goes without saying that one can translate these various vector-valued theorems from the disc to the half-plane, where Lax originally proved these results.

Representations of H^∞

In this section we shall discuss representations (by bounded linear operators on a Hilbert space) of the algebra of bounded analytic functions in the unit disc. The section is brief and the results are rather meager. There are two reasons for including this section. First, as we mentioned earlier, this point of view does lend perspective to some of the results above. Second, the study of representations of H^∞ seems worthy of considerable research.

By a **representation** of H^∞ we mean a mapping $f \to T_f$ from H^∞ into the set of bounded linear operators on some Hilbert space which is an algebra homomorphism, and which carries 1 onto the identity operator. The material above has been concerned with the special class of representations obtained as follows. Choose a Hilbert space K and represent f by the operator "multiplication by f" on the Hilbert space $H^2(K)$:

$$(T_f h)(z) = f(z)h(z).$$

This standard representation of H^∞ on $H^2(K)$ has (of course) many special properties. Some of those which may be of interest from the standpoint of more general representations are as follows:

(i) for every inner function f the operator T_f is an isometry; if $f \neq 1$, then T_f is not unitary on any non-zero subspace;

(ii) $\|f\|_\infty = \|T_f\|$, the representation is isometric;

(iii) if $f_n \to f$ boundedly and pointwise almost everywhere, then $T_{f_n} \to T_f$ in the strong operator topology, i.e., $T_{f_n}(x) \to T_f(x)$ for each vector x.

The results on isometries which we discussed above show that certain of these properties characterize the "standard" representations.

Theorem. *Let* f → T_f *be a representation of* H^∞.

(i) *The representation is unitarily equivalent to the standard representation of* H^∞ *on* H^2 *of some Hilbert space if and only if the operator* T_z *is an isometry which is not unitary on any non-zero subspace.*

(ii) *The representation is unitarily equivalent to the standard representation on* H^2 *of the disc if and only if* T_z *is a non-unitary isometry with no non-trivial completely reducing subspace.*

Proof. Assuming that T_z is isometric and not unitary on a non-zero subspace, we let N be the orthogonal complement of the range of T_z. The underlying Hilbert space then has the form

$$N \oplus T_z(N) \oplus T_z^2(N) \oplus \cdots.$$

In other words, there is an isomorphism of the space onto $H^2(N)$ which carries T_z onto the operator "multiplication by z." We must verify that this isomorphism carries T_f onto multiplication by f, for every f in H^∞. Since T_f is a bounded linear operator, it will suffice to prove this on N; that is, it will suffice to prove that if

$$f(z) = \sum_{k=0}^{\infty} a_k z^k$$

and $n \in N$, then

$$T_f(n) = \sum_{k=0}^{\infty} a_k T_z^k(n).$$

Now

$$T_f(n) - a_0 n = (T_f - a_0 I)n = T_g(n)$$

where $g(z) = f(z) - a_0$. Since $g(0) = 0$, we have $g = zh$ for some h in H^∞. Thus

$$T_g(n) = T_z(T_h(n))$$

proving that $T_g(n)$ is in the range of T_z. It follows that $a_0 n$ is the orthogonal projection of n into the subspace N. By considering $f(z) - a_0 - a_1 z$, the same sort of argument shows that $a_0 n + a_1 T_z(n)$ is the orthogonal projection of $T_f(n)$ into $N \oplus T_z(N)$. Continuing in this way, we prove (i).

Statement (ii) is now merely a repetition of the characterization of the shift operator on H^2 of the disc.

The theorems on the invariant subspaces for the shift operator take the following form, in the language of representations.

Theorem. *Let* K *be a Hilbert space and let* f → M_f *be the standard representation of* H^∞ *on* $H^2(K)$; *i.e.*, M_f *is multiplication by* f. *Let* S *be any subspace of* $H^2(K)$ *which is invariant under this representation. Then the induced representation on* S *is unitarily equivalent to the standard representation of* H^∞ *on* H^2 *of some Hilbert space* N, *where* dim N $\leq$ dim K.

Proof. This is a simple corollary of part (i) of the last theorem.

It seems natural to ask what one can say about a representation $f \to T_f$ of H^∞ under various other sets of hypotheses. For example, what can be said if one assumes only that T_z is an isometry? Not a great deal. For instance, there is a representation of H^∞ with these properties:

(a) $||T_f|| \leq ||f||_\infty$;
(b) for every inner function f the operator T_f is an isometry;
(c) for at least one inner f the isometry T_f is not unitary on any non-zero subspace;
(d) T_z is the identity operator.

If we are willing to add a very special continuity condition to the representation, in addition to the assumption that T_z is isometric, a great deal can be said. This continuity condition is that if $f_n \to f$ boundedly and pointwise almost everywhere, then $T_{f_n} \to T_f$ in the strong operator topology. Suppose T_z is isometric and this continuity prevails. If we let N be the orthogonal complement of the range of T_z, then the underlying space decomposes in the form

$$M \oplus N \oplus T_z(N) \oplus T_z^2(N) \oplus \cdots$$

where M consists of all vectors in the space which are "infinitely divisible" by T_z. Now it is easy to see that M is invariant under all T_f. This does not require continuity of the representation; M is invariant under any operator which commutes with T_z. The continuity guarantees that

$$N \oplus T_z(N) \oplus \cdots$$

is invariant under all T_f; certainly, this space is invariant under T_f for all polynomials f, and the polynomials are dense in H^∞, using the topology of bounded pointwise almost everywhere convergence on the unit circle.

Thus, the representation decomposes into the direct sum of two representations, one of which is equivalent to the standard representation on $H^2(N)$. The other, on the subspace M, is completely determined by the unitary operator U, which we obtain by restricting T_z to M. Of course, U is not an arbitrary unitary operator; but, evidently, U is a unitary operator having the property that if $\{f_n\}$ is a sequence of polynomials which is bounded and converges almost everywhere on the circle, the operators $f_n(U)$ converge in the strong operator topology, i.e., converge pointwise on M. This holds if U is the direct sum of unitary operators, each of which is equivalent to multiplication by $e^{i\theta}$ on $L^2(d\mu)$, where μ is a positive measure on the circle, which is absolutely continuous with respect to Lebesgue measure.

NOTES

The starting point for this chapter is Beurling [8]. See also the paper of Karhunen [51]. In the half-plane, the scalar and vector-valued problems are in Lax [53]. The essence of the generalization to Dirichlet algebras is in Helson-Lowdenslager [43]. The L^2 case and some of its generalizations are in the more recent paper by Helson-Lowdenslager [44]. The relevance of the results on isometries is pointed out by Halmos [39]. His paper contains a more detailed discussion of the vector-valued case. For example, we have not included a uniqueness theorem. Wermer [93] has used the basic result on invariant subspaces to embed "analytic discs" in the space of maximal ideals of a Dirichlet algebra.

EXERCISES

1. Let $\{\alpha_n\}$ be a sequence of points in the open unit such that $\Sigma (1 - |\alpha_n|) < \infty$. Let S be the set of all functions f in H^2 such that $f(\alpha_n) = f'(\alpha_n) = 0$ for each n. Prove that S is a closed subspace of H^2 invariant under multiplication by z. Find the inner function F such that $S = FH^2$.

2. Find a bounded analytic function f in the disc and a closed subspace of H^2 which is invariant under multiplication by f but is not of the form FH^2.

3. Let f be a bounded analytic function in the unit disc. Prove that $1, f, f^2, f^3, \ldots$ form an orthonormal basis for H^2 if and only if $f(z) = \lambda z$, where $|\lambda| = 1$.

4. (Beurling) For f in $H^2, f \neq 0$, define

$$\delta(f) = \exp\left[\log |f(0)| - \frac{1}{2\pi} \int_{-\pi}^{\pi} \log |f(e^{i\theta})| d\theta \right].$$

Prove that $0 \leq \delta(f) \leq 1$ and that δ is multiplicative. Also show that the functions $z^n f$, $n \geq 0$, span H^2 if and only if $\delta(f) = 1$.

5. Let S be the subspace of H^2 consisting of all f such that $f\left(1 - \frac{1}{n^2}\right) = 0$ for all n greater than some positive integer N_f. What is the closure of S?

6. Let f be a square-integrable function on the unit circle. One of the results of this chapter (or Szegö's theorem) implies that the functions $e^{in\theta}f(\theta)$ with $n \geq 0$ span L^2 if and only if the $e^{in\theta}f(\theta)$ with $n \leq 0$ span L^2. Is this obvious, a priori?

7. Which of the following functions f in L^2 of the real line have the property that their right translates span L^2?

(a) $f(x) = \dfrac{1}{1 + x^2}$ (b) $f(x) = \dfrac{1}{1 + ix}$ (c) $f(x) = e^{-|x|}$

8. Let μ be a finite positive Baire measure on the unit circle, and let $H^2(d\mu)$ be the closure in $L^2(d\mu)$ of the polynomials in z. Describe the closed subspaces of $H^2(d\mu)$ which are invariant under multiplication by z. Exercise 3 of Chapter 4 may be of some help.

9. Let F be a non-constant inner function in the disc, and let T be the operator "multiplication by F" on H^2.

(a) Prove that T is an isometry which is unitary on no non-zero subspace.

(b) Prove that T is unitarily equivalent to multiplication by z if and only if

$$F(z) = \lambda \frac{z - \alpha}{1 - \bar{\alpha} z}$$

where $|\lambda| = 1$ and $|\alpha| < 1$.

10. If f is in H^1 of the disc, prove that the functions $z^n f$, $n \geq 0$, span H^1 if and only if f is an outer function.

11. Let $0 < r < 1$, and let T_r be the linear operator on H^2 of the disc defined by

$$(T_r f)(z) = f(rz);$$

that is, T_r is restriction to the disc of radius r. Find all the invariant subspaces for T_r. (Hint: Show that T_r is a positive and completely continuous operator.)

CHAPTER 8

H^p SPACES IN A HALF-PLANE

H^p of the Half-plane

In this chapter we shall be working in the half-plane Re $(w) \geq 0$. If f is analytic in the open right half-plane, we say that f belongs to the class H^p, provided that the L^p norms

$$\int_{-\infty}^{\infty} |f(x+iy)|^p dy$$

are bounded for $x > 0$. We shall establish some of the theory of these spaces. Their study is much more akin to the theory of Fourier transforms than to the theory of Fourier series. We shall work in part with Fourier transforms; however, we shall utilize what we know about H^p of the unit disc to establish some of the fundamentals. This has the disadvantage of being somewhat "unnatural," but it has the advantages of avoiding duplication of proofs and of exhibiting the simple relationship between the H^p spaces of the disc and those of the half-plane.

At the outset, there are a few elementary comments we should make. The conditions imposed on an H^p function in the half-plane are (in a sense) more restrictive than the corresponding conditions in the disc. In order for an analytic f to be in H^p we first must require that the L^p norms of f along vertical lines be finite and then that they be bounded. In the disc there is no question of the finiteness of the L^p norm on the circle of radius r. When we require a bound on the L^p norms on vertical lines, the bound for large x is just as important as the bound for x small, i.e., vertical lines near the boundary. The function

$$f(w) = \frac{1}{1+w} e^w$$

is square-integrable on each vertical line and the integrals

$$\int_{-\infty}^{\infty} |f(x+iy)|^2 dy$$

are bounded on any strip $0 \leq x \leq c$; but these integrals are clearly not bounded as $x \leq \infty$.

The primary tasks for us will be the proof of the existence of boundary values (on the imaginary axis) and the establishment of a Poisson integral formula for recapturing a function from its boundary values. To see what the Poisson kernel is for a half-plane, let us look at the linear fractional map

$$z = \frac{w-1}{w+1}$$

of the half-plane $\operatorname{Re} w > 0$ onto the unit disc $|z| < 1$. On the boundary this map is

$$e^{i\theta} = \frac{it-1}{it+1}$$

from which it is easy to deduce that

$$\frac{1}{2\pi}\frac{d\theta}{dt} = \frac{1}{\pi}(1+t^2)^{-1}.$$

In other words, the normalized Lebesgue measure on the circle corresponds to the Cauchy probability measure

$$\frac{1}{\pi}\frac{dt}{1+t^2}$$

on the imaginary axis. If g is a measurable function on the unit circle, and if

$$f(it) = g(e^{i\theta})$$
$$= g\left(\frac{it-1}{it+1}\right)$$

then g is Lebesgue-integrable if and only if f is integrable with respect to the measure $\frac{1}{1+t^2}dt$. When g is integrable,

$$\frac{1}{2\pi}\int_{-\pi}^{\pi} g(e^{i\theta})d\theta = \frac{1}{\pi}\int_{-\infty}^{\infty} f(it)\cdot \frac{1}{1+t^2}dt.$$

Now it is easy to "lift" the Poisson formula from the disc to the half-plane. The Poisson kernel for the point z in the disc is

$$P_z(\theta) = \operatorname{Re}\left[\frac{e^{i\theta}+z}{e^{i\theta}-z}\right].$$

Since

$$\frac{e^{i\theta}+z}{e^{i\theta}-z} = \frac{\frac{it-1}{it+1}+\frac{w-1}{w+1}}{\frac{it-1}{it+1}-\frac{w-1}{w+1}}$$

$$= \frac{itw-1}{it-w}$$

the Poisson kernel for the point w in the right half-plane should be

$$\operatorname{Re}\left[\frac{itw-1}{it-w}\right] = \frac{x(1+t^2)}{x^2+(y-t)^2}.$$

Of course, we must use this function with the measure $\frac{dt}{\pi(1+t^2)}$, so it is perhaps more sensible to call

$$\frac{1}{1+t^2}\operatorname{Re}\left[\frac{itw-1}{it-w}\right]$$

the **Poisson kernel** for w. From Fatou's theorem in the disc we now have the following.

Theorem. *Let F be a measurable function on the imaginary axis which is integrable with respect to the measure* $(1+t^2)^{-1}dt$. *Define* f *in the right half-plane by*

$$f(x+iy) = \frac{1}{\pi}\int_{-\infty}^{\infty} F(it)\frac{x}{x^2+(y-t)^2}\,dt$$

$$= \frac{1}{\pi}\int_{-\infty}^{\infty} F(it)P_x(y-t)dt.$$

Then f *is harmonic and has non-tangential limits which exist and agree with* F *at almost every point of the imaginary axis.*

Proof. The proof is immediate from Fatou's theorem for the disc and from our computations above. One should remark that non-tangential arcs are preserved, since the map from the disc to the half-plane is conformal.

Several comments are in order. First, note that the above theorem is, in particular, true if F belongs to L^p of Lebesgue measure on the imaginary axis, for some $p \geq 1$. Second, one can, of course, give a direct proof of the above theorem, without appealing to the corresponding result in the disc. Third, if F is the restriction to the imaginary axis of a function analytic in some half-plane $\operatorname{Re}(w) > -\epsilon$, it does not follow that f is analytic. Some control on the analytic function at infinity is necessary to guarantee this. For instance, it is clear that the Poisson formula reproduces any bounded analytic function from its boundary values, but not, say, e^w. Now let us record some of the special properties of f when F is actually in $L^p(dt)$.

Theorem. *Let* $p \geq 1$ *and let* F *be a function in* $L^p(-\infty, \infty)$. *Let* f *be the harmonic function in the right half-plane defined by*

$$f(x+iy) = \frac{1}{\pi}\int_{-\infty}^{\infty} F(t)P_x(y-t)dt.$$

(i) *For each* $x > 0$ *the function* $f_x(y) = f(x + iy)$ *is in* $L^p(-\infty, \infty)$.

(ii) *The L^p-norms* $\|f_x\|_p$ *are bounded for* $x > 0$. *In fact,* $\|f_x\|_p$ *is a decreasing function of* x *for* $x > 0$.

(iii) *The functions f_x converge to F in* $L^p(-\infty, \infty)$ *as* $x \to 0$.

(iv) $f(w)$ *tends uniformly to zero as w tends to infinity inside any fixed half-plane* Re $(w) \geq \delta > 0$.

Proof. We are not on a finite measure space, so F need not be integrable. But since $\frac{1}{\pi} P_x$ is integrable, its convolution with F is a function f_x in L^p. Also,

$$\|G*F\|_p \leq \|G\|_1 \|F\|_p$$

and since $P_x \geq 0$ and

$$\frac{1}{\pi} \int_{-\infty}^{\infty} P_x(t)dt = 1,$$

we have

$$\|f_x\|_p \leq \|F\|_p.$$

To see that $\|f_x\|_p$ is a decreasing function of x, suppose $x_1 < x_2$. Verify that

$$f_{x_2}(y) = \frac{1}{\pi} \int_{-\infty}^{\infty} f_{x_1}(t) P_{x_2-x_1}(y-t) dt$$

and repeat the above argument, replacing F by f_{x_1} and P_x by $P_{x_2-x_1}$. The convergence of f_x to F in L^p is a standard approximate identity argument, like several we have given.

To prove (iv), observe that since $\frac{1}{\pi} P_x(y-t) dt$ is a positive measure of mass 1 we have

$$|f(x+iy)|^p \leq \frac{1}{\pi} \int_{-\infty}^{\infty} |F(t)|^p P_x(y-t) dt.$$

Suppose $\epsilon > 0$. Since F is in L^p, we can choose $T > 0$ such that

$$\int_{-\infty}^{-T} |F|^p dt + \int_{T}^{\infty} |F|^p dt < \epsilon.$$

Now $P_x(y-t) \leq \frac{1}{x}$, and if we keep $x \geq \delta > 0$, we shall have

$$|f(x+iy)|^p \leq \frac{1}{\pi} \int_{-T}^{T} |F(t)|^p \frac{x}{x^2 + (y-t)^2} dt + \frac{\epsilon}{\pi \delta}.$$

Now with T fixed, the last integral is obviously $0(|x+iy|^{-1})$ in $x \geq \delta$.

Boundary Values for H^p Functions

We are now going to prove that if f is in H^p of the right half-plane for some $p \geq 1$, then f has non-tangential limits at almost every point of the imaginary axis, the boundary value function F is in L^p, and f is the Poisson integral of F. As we shall see, this follows rather easily from the fact that such an f possesses property (iv) of the last theorem.

Theorem. *Let f be in* H^p *of the right half-plane for some* $p \geq 1$. *Then* f(w) *tends uniformly to zero as w tends to infinity inside any fixed half-plane* Re (w) $\geq \delta > 0$.

Proof. The proof will proceed by reducing to the last theorem. That is, we shall show that for Re $(w) > x_0 > 0$ we can write f as the Poisson integral of its values on the line Re $(w) = x_0$. Fix w with Re $(w) > 0$. Choose positive numbers x_0, x, and y such that

$$x_0 < \text{Re}(w) < x$$
$$-y < \text{Im}(w) < y.$$

We represent f by means of the Cauchy integral formula applied to the rectangle with vertices $x_0 \pm iy$, $x \pm iy$. We obtain

$$2\pi i f(w) = i \int_{-y}^{y} \frac{f(x+it)}{x+it-w} dt - i \int_{-y}^{y} \frac{f(x_0+it)}{x_0+it-w} dt$$

$$+ \int_{x_0}^{x} \frac{f(s-iy)}{s-iy-w} ds - \int_{x_0}^{x} \frac{f(s+iy)}{s+iy-w} ds$$

$$= I_1(x; y) - I_1(x_0; y) + I_2(y) - I_2(-y).$$

Choose a large positive number Y, say $Y > 2|\text{Im } w|$. We shall average the above expression for $f(w)$ over the interval $(Y, 2Y)$. First, note that

$$\frac{1}{Y} \int_Y^{2Y} |I_2(\pm y)| dy \leq \frac{1}{Y} \int_Y^{2Y} dy \cdot \int_{x_0}^{x} \frac{|f(s \pm iy)|}{|s+iy-w|} ds$$

$$\leq \frac{2}{Y^2} \int_{x_0}^{x} \int_Y^{2Y} |f(s \pm iy)| dy ds$$

since $|s \pm iy - w| \geq Y/2$ in the range of y's considered. Now, we estimate the inside integral by Hölder's inequality, using the fact that there is a fixed M such that

$$\int_{-\infty}^{\infty} |f(s+iy)|^p dy \leq M^p.$$

If q is the conjugate to p, i.e., if $\dfrac{1}{p} + \dfrac{1}{q} = 1$, this estimate gives

$$\frac{1}{Y} \int_Y^{2Y} |I_2(\pm iy)| dy \leq \frac{2}{Y^2} \int_{x_0}^{x} ds \cdot Y^{1/q} \cdot M$$

$$= 2M(x - x_0) \cdot Y^{1/q-2}$$

$$\to 0 \quad \text{as } Y \to \infty.$$

Averaging $2\pi i f(w)$, we obtain

$$2\pi i f(w) = \lim_{Y\to\infty} \frac{1}{Y} \int_Y^{2Y} [I_1(x;y) - I_1(x_0;y)]dy.$$

Now it is easy to see that

$$\lim_{Y\to\infty} \frac{1}{Y} \int_Y^{2Y} I_1(x;y)dy = i \int_{-\infty}^{\infty} \frac{f(x+it)}{x+it-w} dt = I_1(x;\infty)$$

$$\lim_{Y\to\infty} \frac{1}{Y} \int_Y^{2Y} I_1(x_0;y)dy = i \int_{-\infty}^{\infty} \frac{f(x_0+it)}{x_0+it-w} dt = I_1(x_0;\infty).$$

This is an immediate consequence of the fact that

$$\int_{-\infty}^{\infty} \frac{|f(s+it)|}{|s+it-w|} dt < \infty.$$

We obtain, therefore,

$$2\pi i f(w) = I_1(x;\infty) - I_1(x_0;\infty).$$

We shall now let $x \to \infty$. For $p = 1$,

$$|I_1(x;\infty)| \leq \int_{-\infty}^{\infty} \frac{|f(x+it)|}{x - \operatorname{Re} w} dt \leq M \cdot (x - \operatorname{Re} w) = O\left(\frac{1}{x}\right)$$

and for $p > 1$,

$$|I_1(x;\infty)| \leq M \cdot \left[\int_{-\infty}^{\infty} \frac{dt}{|x+it-w|^q}\right]^{1/q} = O(x^{-1+1/q})$$

so $I_1(x;\infty) \to 0$ as $x \to \infty$. Thus we obtain

$$2\pi i f(w) = -i \int_{-\infty}^{\infty} \frac{f(x_0+it)}{x_0+it-w} dt$$

for any x_0 which satisfies $0 < x_0 < \operatorname{Re} w$.

We have a Cauchy integral representation for f, from which we can easily obtain a Poisson integral representation. Let

$$w = \xi + i\eta \quad \text{and} \quad w' = 2x_0 - \xi + i\eta.$$

Now $\operatorname{Re} w' < x_0$, so if we use the Cauchy integral theorem and repeat the above argument, we obtain

$$0 = \int_{-\infty}^{\infty} \frac{f(x_0+it)}{x_0+it-w'} dt.$$

If we subtract this from our Cauchy integral for $f(w)$ and remember that $w - w' = 2(\xi - x_0)$, we obtain

$$f(\xi+i\eta) = \frac{1}{\pi} \int_{-\infty}^{\infty} f(x_0+it) \frac{(x_0-\xi)}{(x_0-\xi)^2 + (t-\eta)^2} dt, \quad 0 < x_0 < \xi.$$

By part (iv) of the last theorem we have immediately that $f(w)$ tends uniformly to zero as w tends to infinity inside any fixed half-plane $\operatorname{Re}(w) \geq \delta > 0$.

Theorem. *Let* $p \geq 1$ *and let* f *be any function in* H^p *of the right half-plane. Let* g *be the function in the unit disc defined by*

$$g(z) = f\left(\frac{1+z}{1-z}\right).$$

Then g *is in* H^p *of the disc.*

Proof. Let C_r be the circle of radius r in the disc: $C_r = \{|z| = r\}$. Then we wish to bound the integrals

$$\int_{-\pi}^{\pi} |g(re^{i\theta})|^p d\theta = \frac{1}{r} \int_{C_r} |g(z)|^p |dz|$$

as $r \to 1$.

For each $\delta > 0$, the line $x = \delta$ in the right half-plane is transformed by our linear fractional map into a circle Γ_δ in the disc, which is tangent to the unit circle at $z = 1$. Given $r < 1$, we can choose a $\delta > 0$ which is sufficiently small that C_r lies in the interior of the circle Γ_δ. By the last theorem, the function f is continuous in the half-plane Re $(w) \geq \delta$, including the point at infinity. This means that g is analytic in the interior of the circle Γ_δ and is continuous on the closed disc bounded by Γ_δ. Since C_r is inside this disc, it follows that

$$\int_{C_r} |g(z)|^p |dz| \leq 2 \int_{\Gamma_\delta} |g(z)|^p |dz|.$$

We shall comment on this below. Assuming this inequality for now, we obtain

$$\int_{-\pi}^{\pi} |g(re^{i\theta})|^p d\theta \leq \frac{2}{r} \int_{\Gamma_\delta} |g(z)|^p |dz|$$

$$= \frac{2}{r} \int_{\text{Re } w = \delta} |f(w)|^p \frac{2|dw|}{|1+w|^2}$$

$$= \frac{4}{r} \int_{-\infty}^{\infty} |f(\delta + it)|^p \frac{dt}{(1+\delta)^2 + t^2}$$

$$\leq \frac{4}{r} \int_{-\infty}^{\infty} |f(\delta + it)|^p dt$$

$$\leq \frac{4}{r} M^p$$

where M is the bound for the L^p norms of f on vertical lines. We conclude that g is in H^p of the disc.

We used an inequality in the above proof, which can be stated as follows. Suppose g is analytic inside the circle Γ and continuous on the closed disc bounded by Γ. If C is a circle interior to that disc, then

$$\int_C |g(z)|^p |dz| \leq 2 \int_\Gamma |g(z)|^p |dz|.$$

To prove this, one may assume that Γ is the unit circle. The result then states that if g is continuous on the closed unit disc and analytic in the interior, and if C is a circle in the open unit disc, then

$$\int_C |g(z)|^p |dz| \leq 2 \int_{-\pi}^{\pi} |g(e^{i\theta})|^p d\theta.$$

If one represents g as the Poisson integral of its boundary values, this is trivial to verify.

Now we have the theorem we have been seeking.

Theorem. *Let $p \geq 1$ and let f be a function in H^p of the right half-plane.*
  (i) *f has non-tangential limits at almost every point of the imaginary axis.*
  (ii) *The boundary values of f are in L^p and*

$$f(x + iy) = \frac{1}{\pi} \int_{-\infty}^{\infty} f(it) \frac{x}{x^2 + (y - t)^2} dt, \quad x > 0.$$

  (iii) *The functions $f_x(y) = f(x + iy)$ converge in L^p-norm to $f(iy)$ as $x \to 0$.*

Proof. Let g be the function in the unit disc defined by

$$g(z) = f\left(\frac{1 + z}{1 - z}\right).$$

By the last theorem, g is in H^p of the disc. Thus g has non-tangential limits at almost every point of the unit circle, and g is the Poisson integral of its boundary values:

$$g(z) = \frac{1}{2\pi} \int_{-\pi}^{\pi} g(e^{i\theta}) P_z(\theta) d\theta.$$

Since the linear fractional map from the disc to the half-plane is conformal, we deduce immediately that f has non-tangential limits at almost every point of the imaginary axis. As we observed at the beginning of this chapter, the boundary values of f are in $L^p\left(\frac{dt}{1 + t^2}\right)$, and f is represented by the Poisson formula of (ii). Of course, the boundary values of f are actually in $L^p(dt)$. By Fatou's lemma

$$\int_{-\infty}^{\infty} |f(it)|^p dt \leq \varlimsup_{x \to 0} \int_{-\infty}^{\infty} |f(x + it)|^p dt$$

$$\leq M^p.$$

This proves (i) and (ii). Statement (iii) follows from (ii), as we observed in a previous theorem. We have stated it again for emphasis.

The Relation between the H^p Spaces for the Disc and Half-plane

The proof of the last theorem made essential use of the fact that each H^p function f in the right half-plane is transformed by the linear fractional map

$$w = \frac{1+z}{1-z}$$

into a function g in H^p of the disc:

$$g(z) = f\left(\frac{1+z}{1-z}\right).$$

We have, therefore, a mapping from H^p of the half-plane to a subspace of H^p of the disc. It is quite easy to determine what this subspace is. It consists of all functions g in H^p of the disc such that

$$(1-z)^{-2/p} g(z)$$

is also in H^p of the disc. If g is in H^p of the disc, the corresponding analytic f in the right half-plane need not be in H^p of the half-plane. All that one can state is that f is analytic for Re $(w) > 0$ and is the Poisson integral of a function on the imaginary axis which is in L^p of the measure $(1 + t^2)^{-1} dt$. In other words,

$$\frac{|f(it)|^p}{1+t^2}$$

is Lebesgue integrable. Suppose we select an analytic branch of the logarithm of $1 + w$ in the right half-plane. Then

$$\frac{f(w)}{(1+w)^{2/p}}$$

is in L^p on the imaginary axis. Thus, f is in H^p of the right half-plane if and only if $(1+w)^{2/p} f(w)$ is in $L^p(dt/1+t^2)$ on the imaginary axis (if we assume, as always, that g is in H^p of the disc). Since

$$1 + w = \frac{2}{1-z}$$

we see that f is in H^p of the half-plane exactly when

$$\frac{g(z)}{(1-z)^{2/p}}$$

is in L^p on the unit circle. Now we need the following.

Lemma. *Let* $p \geq 1$, *let* $\alpha > 0$, *and let* g *be a function in* H^p *of the disc. A necessary and sufficient condition that the function*

$$h(z) = \frac{g(z)}{(1-z)^\alpha}$$

be in H^p *is that* $h(e^{i\theta})$ *be in* L^p *on the unit circle.*

Proof. Now

$$|g(z)|^p \leq \exp\left[\frac{1}{2\pi}\int_{-\pi}^{\pi} \log |g(e^{i\theta})|^p P_z(\theta) d\theta\right]$$

$$|1-z| = \exp\left[\frac{1}{2\pi}\int_{-\pi}^{\pi} \log |1-e^{i\theta}| P_z(\theta) d\theta\right]$$

so that

$$|h(z)|^p \leq \exp\left[\frac{1}{2\pi}\int_{-\pi}^{\pi} P_z(\theta) \log |h(e^{i\theta})|^p d\theta\right]$$

$$\leq \frac{1}{2\pi}\int_{-\pi}^{\pi} P_z(\theta) |h(e^{i\theta})|^p d\theta.$$

Thus

$$\int_{-\pi}^{\pi} |h(re^{it})|^p dt \leq \int_{-\pi}^{\pi} |h(e^{i\theta})|^p d\theta.$$

We should point out that the only property of $(1-z)^\alpha$ which is relevant in the lemma is that it is an outer function. Returning to our g before the lemma, we see that g transforms into a function in H^p of the half-plane if and only if

$$(1-z)^{-2/p} g(z)$$

is also in H^p of the disc. What we have proved may be summarized as follows.

Theorem. *Let g be an analytic function in the unit disc, and let f be the analytic function in the right half-plane defined by*

$$f(w) = g\left(\frac{w-1}{w+1}\right).$$

If $p \geq 1$*, then f is in H^p of the right half-plane if and only if*

$$g(z) = (1-z)^{2/p} G(z)$$

where G is in H^p of the unit disc. Equivalently, g is in H^p of the unit disc if and only if

$$f(w) = (1+w)^{2/p} F(w)$$

where F is in H^p of the right half-plane.

There are some remarks we might make. The content of the last theorem is, perhaps, more clear if we state it roughly. The relation between the two H^p spaces is this. Start with H^p of the disc; use the linear fractional map to transform this space to a space $\tilde{H}^p$ of functions in the right half-plane; choose an analytic branch of $(1+w)^{2/p}$ on Re $w > 0$; divide every function in $\tilde{H}^p$ by $(1+w)^{2/p}$; the resulting space is H^p of the half-plane. Indeed, up to a constant, this map is a Banach space isometry between the two H^p spaces. If

then
$$h(w) = (1+w)^{-2/p} g\left(\frac{w-1}{w+1}\right)$$

$$\frac{1}{\pi}\int_{-\infty}^{\infty} |h(it)|^p dt = \frac{1}{2\pi}\int_{-\pi}^{\pi} |g(e^{i\theta})|^p d\theta.$$

Since $\tilde{H}^p$ consists of all analytic f in the right half-plane which are the Poisson integral of a function on the imaginary axis in $L^p(dt/1+t^2)$, and H^p consists of all analytic f which are the Poisson integral of a function on the imaginary axis in $L^p(dt)$, the relation

$$\tilde{H}^p = (1+w)^{2/p} H^p$$

seems evident. On the boundary the relationship is clearly right. The only point to be checked is that after multiplying an H^p function by $(1+w)^{2/p}$ the resulting analytic function is still the Poisson integral of its boundary values. This is what we did in the lemma above.

For the case $p = 2$ we have the relation $\tilde{H}^2 = (1+w)H^2$ which we discussed in the last chapter.

The Paley-Wiener Theorem

If one uses the Plancherel theorem, one obtains from our results here the one-sided Paley-Wiener theorem.

Theorem (Paley-Wiener). *A complex-valued function* f *in the right half-plane belongs to the class* H² *if and only if* f *has the form*

$$f(w) = \frac{1}{\sqrt{2\pi}} \int_0^\infty \hat{f}(t) e^{-wt} dt$$

for some function f̂ *in* L²(0, ∞). *This representation is unique.*

Proof. The function $\hat{f}$ will be defined by

$$\hat{f}(x) = \frac{1}{\sqrt{2\pi}} \int_{-\infty}^\infty f(it) e^{ixt} dt.$$

Literally, this definition makes sense only if f is integrable. The Plancherel theorem asserts that if f is in $L^1 \cap L^2$ then $\hat{f}$ is in L^2 and $||f||_2 = ||\hat{f}||_2$, the set of such $\hat{f}$ is dense in L^2, and so the Fourier transform $f \to \hat{f}$ can be uniquely extended to a unitary mapping of L^2 onto L^2. This defines the Fourier transform $\hat{f}$ for any f in L^2. We represent $\hat{f}$ by the above formula, with the understanding that this is correct if f is in L^1, but must be interpreted in a limiting sense in general. With the same convention about limits, we also have the inversion formula

$$f(it) = \frac{1}{\sqrt{2\pi}} \int_{-\infty}^\infty \hat{f}(x) e^{-ixt} dx.$$

Now it is quite easy to verify that if $\hat{f}$ is in L^2 and vanishes on the left half-line, the corresponding f is the boundary function of an H^2 function. Just extend f to the half-plane as in the statement of the theorem.

We want to prove that if f is in H^2 the Fourier transform $\hat{f}$ vanishes on the left half-line. First, let us observe that it will suffice to prove this when f is in H^1, for $H^1 \cap H^2$ is plainly dense in H^2. If f is in H^2, f is the limit in L^2 norm of the functions

$$\left[1 - \left(\frac{w}{w+1}\right)^n\right] f(w)$$

each of which is in $H^1 \cap H^2$.

Suppose f is in H^1 and $x < 0$. Let

$$h(w) = (1 - w^2)e^{xw}f(w)$$

$$= \frac{1-w}{1+w} e^{xw}(1+w)^2 f(w).$$

Now $(1 + w)^2 f(w)$ is in what we called $\tilde{H}^1$, the image of H^1 of the disc. Evidently, h is also in this space, because $\frac{1-w}{1+w}$ is bounded on $\operatorname{Re}(w) \geq 0$, and for $x < 0$ the function e^{xw} is also bounded on that half-plane. But then

$$\frac{1}{\pi}\int_{-\infty}^{\infty} f(it) e^{ixt} dt = \frac{1}{\pi}\int_{-\infty}^{\infty} h(it) \cdot \frac{1}{1+t^2} dt$$

$$= h(1)$$

$$= 0.$$

Here we have used the fact that $[\pi(1 + t^2)]^{-1}$ is the Poisson kernel for the point $w = 1$. That completes the proof.

Factorization for H^p Functions in a Half-plane

By direct translation of the corresponding results in the unit disc we obtain the following results about a non-zero function f in H^p of the right half-plane, $p \geq 1$.

(i) If $\beta_1, \beta_2, \ldots$ are the zeros different from 1 of f in $\operatorname{Re}(w) > 0$, then

$$\sum_n \frac{\operatorname{Re}(\beta_n)}{1 + |\beta_n|^2} < \infty$$

which is the necessary and sufficient condition for the convergence of the Blaschke product

$$B(w) = \left(\frac{w-1}{w+1}\right)^k \prod_n \frac{|1 - \beta_n^2|}{1 - \beta_n^2} \cdot \frac{w - \beta_n}{w + \bar{\beta}_n}.$$

Of course, k denotes the order of the zero of f at $w = 1$ and

$$g(w) = \frac{f(w)}{B(w)}$$

is a function in H^p without zeros.

(ii) The boundary values of f cannot vanish on a set of positive Lebesgue measure; indeed,

$$\int_{-\infty}^{\infty} \log |f(it)| \cdot \frac{1}{1+t^2} dt > -\infty.$$

The function

$$F(w) = \exp\left[\frac{1}{\pi}\int_{-\infty}^{\infty} \log |f(it)| \frac{tw+i}{t+iw}\frac{dt}{1+t^2}\right]$$

is in H^p; also $|F| = |f|$ almost everywhere on the imaginary axis and $|F(w)| \geq |f(w)|$ on the half-plane.

(iii) If $\lambda = e^{ia}$, where $a = \arg\left(\frac{f}{B}\right)(1)$, then the function

$$S(w) = \frac{f(w)}{\lambda B(w) F(w)}$$

is uniquely representable in the form

$$S(w) = e^{-\rho w} \exp\left[-\int \frac{tw+i}{t+iw} d\mu(t)\right]$$

where μ is a finite singular positive measure on the imaginary axis and ρ is a non-negative real number.

NOTES

The basic results on H^p of the half-plane are due to Paley and Wiener [67] for $p = 2$ and to Bochner [12] for $p = 1$. The arguments used here are basically those of Hille and Tamarkin [46] which are similar to the Paley-Wiener methods, and to those of Gabriel [29]. Other approaches can be used, particularly ones which use more about Fourier transforms and less about the disc; e.g., see Titchmarsh [88]. It seems difficult to find the relationship between H^p of the disc and H^p of the half-plane stated anywhere. For a discussion of Fourier transforms, the Plancherel theorem, etc., see the books by Paley and Wiener [68], Wiener [95], and Titchmarsh [88]. For results on factorization of functions analytic in a half-plane see Gabriel [29, 30]. The inequality about $\int |f|^p |dw|$ along curves is discussed in Gabriel [31] and Carlson [19] for subharmonic functions.

EXERCISES

1. Let f be an analytic function in the right half-plane and let g be the function in the unit disc:

$$g(z) = f\left(\frac{1+z}{1-z}\right).$$

We consider the integrals

$$\int_{-\infty}^{\infty} |f(x+iy)|^2 \cdot \frac{1}{1+y^2} dy, \quad x > 0.$$

(a) If these integrals are bounded, show that g is in H^2 of the disc.

(b) Exhibit a g in H^2 so that the integrals are not bounded.

2. Let f be analytic in the half-plane Re $(w) > 0$. Show that f is in H^2 if and only if there exists a positive number δ such that

(i) the integrals

$$\int_{-\infty}^{\infty} |f(x+iy)|^2 dy$$

are bounded for $0 < x < \delta$

(ii) f is bounded on the half-plane Re $(w) \geq \delta$.

3. For f in H^2 of the right half-plane, define

$$(Uf)(w) = \frac{1}{w} f\left(\frac{1}{w}\right).$$

Show that U is a unitary mapping of H^2 onto H^2. Replace inversion by a conformal map of the half-plane onto itself and prove the same result, thereby obtaining a group of unitary operators on H^2.

4. Let f be analytic in the half-plane Re $(w) > 0$. Prove that f is in H^1 if and only if $f = gh$, where g and h are in H^2 of the right half-plane.

5. (Akutowitz [3]). Let f be analytic in the half-plane Re $(w) > 0$. Prove that f is a Blaschke product if and only if

(a) $|f(w)| \leq 1$.

(b) $\lim_{x \to 0} \int_{-\infty}^{\infty} \log |f(x+iy)| \frac{dy}{1+y^2} = 0.$

6. Let f be a *non-negative* harmonic function in the right half-plane. Prove the following.

(a) $f(x+iy) = \rho x + \frac{1}{\pi} \int \frac{x}{x^2 + (y-t)^2} d\mu(t)$, where ρ is a non-negative constant and μ is a finite positive measure on the imaginary axis.

(b) $\int_{-\infty}^{\infty} \frac{f(x+iy)}{1+y^2} dy < \infty$ for all $x > 0$.

(c) (Loomis-Widder [55]) If f is harmonic on $x \geq 0$ and vanishes on $x = 0$, then $f(x+iy) = \rho x$.

7. Let $a > 0$. Let T_a be the linear operator on H^2 of the right half-plane defined by $(T_a f)(w) = f(a+w)$. Use the Paley-Wiener theorem to describe the invariant subspaces for T_a.

8. Let $g \in L^1(-\infty, \infty)$. If $f(iy)$ is the boundary function of an H^2 function in the right half-plane, define

$$(T_g f)(iy) = \int_{-\infty}^{\infty} g(y-t) f(it) dt.$$

Prove that T_g is a bounded linear operator on H^2 (regarded as a subspace of L^2 of the imaginary axis). Let $a > 0$, and let T_a be the operator of Exercise 7. Prove that T_a is one of the operators T_g and that any subspace of H^2 invariant under T_a is also invariant under every T_g.

9. Does the operator $(Tf)(w) = f(\frac{1}{2} w)$ map H^2 of the right half-plane into itself? If so, is T a bounded operator?

10. Prove that every function f in L^2 of the real line is uniquely expressible in the form $f = f_+ + f_-$, where f_+ is in H^2 of the upper half-plane and f_- is in H^2 of the lower half-plane. Show that

$$f_+(w) = \frac{1}{2\pi i} \int_{-\infty}^{\infty} \frac{f(x)}{x - w} dx, \quad \text{Re}(w) > 0$$

and a similar formula defines f_- for $\text{Re}(w) < 0$.

11. Refer to Exercise 10 and let H^+ be the set of L^2 functions on the line which are boundary values of functions in H^2 of the upper half-plane, and define H^- similarly, using the lower half-plane. Show that $L^2 = H^+ \oplus H^-$ (orthogonally), and that both H^+ and H^- are translation invariant. Find all orthogonal direct sum decompositions $L^2 = S \oplus T$, where S and T are (closed) translation invariant subspaces.

CHAPTER 9

H^p AS A BANACH SPACE

The purpose of this chapter is to relate some of the work of recent years on the Banach space structure of the H^p spaces in the unit disc ($p \geq 1$). As usual, we will deal with H^p either as the space of analytic functions in the unit disc for which the functions $f_r(\theta) = f(re^{i\theta})$ are bounded in L^p norm, or as the space of L^p functions f on the unit circle for which

$$\int_{-\pi}^{\pi} e^{in\theta} f(\theta) d\theta = 0, \quad n = 1, 2, 3, \ldots.$$

Described in the latter way, H^p is easily seen to be a commutative Banach algebra, using convolution as multiplication; however, a discussion of this aspect of the spaces would take us too far afield, so we shall be primarily concerned with the linear structure. As it happens, H^∞ is also a commutative Banach algebra under pointwise multiplication. We shall discuss this Banach algebra in some detail in the next chapter, and will treat it in a limited sense in this chapter, since the algebraic structure of H^∞ is of considerable aid in studying H^∞ and H^1 as Banach spaces. The material we cover here is as follows: (1) the characterization of the extreme points in the unit ball of H^p, (2) the isometries of H^∞ and H^1, and (3) projections of L^p onto H^p.

Extreme Points

In studying a Banach space X, one problem of considerable interest is that of describing the geometry of the unit ball,

$$\Sigma = \{x \in X; \|x\| \leq 1\}.$$

In particular, one would like to find the extreme points of Σ, that is, the points in Σ which are not a proper convex combination of two distinct points of Σ. If X is a Hilbert space, the extreme points of Σ are precisely those on the unit sphere, i.e., those of norm 1. This is also true of the L^p spaces for $1 < p < \infty$. On the other hand, the unit ball in L^1 has no extreme points whatsoever, and one must presume that this is somewhat typical for a Banach space. Certainly, then, a description of the extreme points of Σ does not convey complete information about the geometry of Σ. Still, such a description can be very useful.

We are interested in the extreme points of the unit ball in H^p, $1 \leq p \leq \infty$. For $1 < p < \infty$ this presents no problem, for, as we mentioned above, every point on the unit sphere of L^p is an extreme point of the unit ball in L^p. Obviously, then, the extreme points in the unit ball of H^p are exactly the functions of norm 1 ($1 < p < \infty$). This leaves us the two cases $p = 1$ and $p = \infty$. In the case $p = 1$ it is not immediately evident that the unit ball should have any extreme points, since the unit ball in L^1 has none. But H^1 differs radically from L^1 in that it is the conjugate space of a Banach space. A general theorem of Krein and Milman states that if a Banach space X is (isometrically isomorphic to) the conjugate space of some Banach space Y, then the unit ball in X not only has extreme points, but has sufficiently many to "span" the unit ball. The form of this theorem which is relevant here is the following.

Theorem (Krein-Milman). *Let Y be a Banach space, and let K be a non-empty convex subset of the conjugate space Y^*. Suppose that K is compact in the weak-star topology on Y^*. Then K has an extreme point; in fact, K is the weak-star closure of the convex hull of its extreme points.*

The theorem applies in particular to the case when K is the unit ball in Y^*. At one point later on, we shall have need of this theorem, and we shall use it without presenting a proof; however, most of what we do in this chapter does not depend upon this result. Our reason for stating it now is to suggest that the unit balls in H^1 and H^∞ should have "quite a few" extreme points. Of course, in order that the theorem be applicable, we need the following.

Theorem. *Every H^p space ($1 \leq p \leq \infty$) is (isometrically isomorphic to) the conjugate space of a Banach space.*

Proof. For $1 < p \leq \infty$, this result is more or less evident from the corresponding fact about L^p. As a space of functions on the unit circle, H^p is defined as the set of all f in L^p which annihilate $e^{in\theta}$ for $n \geq 1$. Since L^p is the conjugate space of L^q, where $\dfrac{1}{p} + \dfrac{1}{q} = 1$, we see that H^p is the annihilator of the closed subspace of L^q which is spanned by the functions $e^{in\theta}$, $n = 1, 2, 3, \ldots$. The latter space is H_0^q, the space of H^q functions which vanish at the origin. From this it is easy to identify H^p with the conjugate space of the quotient space L^q/H_0^q. In particular, H^∞ "is" the conjugate space of the quotient L^1/H_0^1. The fact that H^1 is a conjugate space is not so evident. This is essentially the content of the F. and M. Riesz theorem on measures: a measure μ on the circle which annihilates $e^{in\theta}$ for $n \geq 1$ is absolutely continuous with respect to Lebesgue measure; i.e., it has the form $d\mu = h d\theta$ with h in H^1. Using this result and the fact that the space of finite complex measures on the circle is the conjugate space of the space C of continuous functions, one can proceed as above to identify H^1 with the conjugate space of C/A_0, where A_0 is the uniform closure of the polynomials vanishing at the origin.

Before taking up the extreme points for H^1 and H^∞, let us make a few elementary observations about the unit ball Σ in the Banach space X. First, the statement that x is an extreme point of Σ simply means that if $x = \frac{1}{2}(y+z)$ with y and z in Σ, then $x = y = z$. For if

$$x = \lambda a + (1-\lambda)b$$

where a and b are distinct elements in Σ and $0 < \lambda < 1$, then we can choose two distinct points y and z on the line segment from b to a such that $x = \frac{1}{2}(y+z)$. Second, any extreme point of Σ necessarily has norm 1; for if $\|x\| < 1$, we can write $x = \frac{1}{2}(y+z)$, where y and z are distinct scalar multiples of x, each having norm less than 1. Third, the condition that x be an extreme point of Σ can also be phrased as follows: if t is any element of X such that $\|x+t\| \leq 1$ and $\|x-t\| \leq 1$, then $t = 0$. For if $x = \frac{1}{2}(y+z)$ with y and z in Σ, put $t = x - y$ so that

$$z = x + t \quad \text{and} \quad \|x+t\| \leq 1, \quad \|x-t\| \leq 1.$$

The condition $x = y = z$ is equivalent to $t = 0$.

Now let us consider H^∞, the space of bounded analytic functions in the unit disc. Since the unit ball of H^∞ consists of those functions which map the disc into the disc, one feels that the extreme functions should be those which take extreme values fairly often. In L^∞ this is the complete answer: the extreme functions in the ball are those which are of modulus 1 almost everywhere, i.e., those whose values are extreme points of the disc. Clearly, then, any f in H^∞ which is of modulus 1 on the unit circle (any inner f) will be extreme in the unit ball of H^∞. However, there are additional extreme points. For example, suppose f is in H^∞, $|f| \leq 1$, and $|f| = 1$ on any set E of positive measure on the circle. Then f is an extreme point of the ball in H^∞; for if $f = \frac{1}{2}(g+h)$ where $|g| \leq 1$ and $|h| \leq 1$ we must have $f = g = h$ on the set E. Since the measure of E is positive while f, g, and h are analytic, this implies $f = g = h$. Neither do these functions exhaust the extreme points. The complete answer is the following.

Theorem. *A function* f *in* H^∞ *is an extreme point of the unit ball in* H^∞ *if and only if* $|f(z)| \leq 1$ *and*

$$\int_{-\pi}^{\pi} \log[1 - |f(e^{i\theta})|]d\theta = -\infty.$$

Proof. Suppose the integral condition holds. Let g be any function in H^∞ such that $\|f+g\|_\infty \leq 1$ and $\|f-g\|_\infty \leq 1$. Then

$$|f(z)|^2 + |g(z)|^2 \leq 1.$$

Therefore
$$|g(e^{i\theta})|^2 \leq 1 - |f(e^{i\theta})|^2$$
and so
$$2\int_{-\pi}^{\pi} \log |g(e^{i\theta})|d\theta \leq 2\pi \log 2 + \int_{-\pi}^{\pi} \log [1 - |f(e^{i\theta})|]d\theta$$
$$= -\infty.$$

This implies that $g = 0$. We conclude that f is extreme.

Now suppose the integral condition on f fails to hold, i.e., that $\log (1 - |f(e^{i\theta})|)$ is integrable. Let
$$g(z) = \exp\left[\frac{1}{2\pi}\int_{-\pi}^{\pi} \frac{e^{i\theta} + z}{e^{i\theta} - z} \log (1 - |f(e^{i\theta})|)d\theta\right].$$
Then $g \neq 0$ and g is a bounded analytic function; indeed,
$$|g(e^{i\theta})| = 1 - |f(e^{i\theta})|.$$
Then $\|f + g\|_\infty \leq 1$ and $\|f - g\|_\infty \leq 1$, so f is not extreme.

We should mention that this result also holds for the Banach space A of continuous functions on the closed disc which are analytic in the interior. The first half of the proof carries over as above, because if f is extreme in the unit ball of H^∞, it is certainly extreme in the unit ball of the subspace A. For the second half of the proof, we must arrange that g have continuous boundary values. If f is in A and $\log (1 - |f(e^{i\theta})|)$ is integrable, we can choose a continuous function u on the unit circle such that $0 \leq u \leq 1 - |f|$, $\log u$ is integrable, and u is continuously differentiable on each open arc of the set where $|f| \neq 1$. If we then put
$$g(z) = \exp\left[\frac{1}{2\pi}\int_{-\pi}^{\pi} \frac{e^{i\theta} + z}{e^{i\theta} - z} \log u\, d\theta\right]$$
g will have continuous boundary values and will satisfy
$$|g(e^{i\theta})| \leq 1 - |f(e^{i\theta})|.$$
Again, f is not extreme.

Now we turn to the more interesting problem of the extreme points in the unit ball of H^1. Here we cannot be motivated by the corresponding result for L^1, since the unit ball in L^1 has no extreme points. We must search for a class of functions in H^1 possessing an extremal property. The relevant class here is that comprised of the outer functions of norm 1. An outer function f in H^1 is characterized by an extremal property: if g is in H^1 and $|f| \geq |g|$ on the circle, then $|f(z)| \geq |g(z)|$ for each z in the disc. So the following result is not too surprising.

Theorem (Rudin-de Leeuw). *Let* f *be a function in* H^1. *Then* f *is an extreme point of the unit ball in* H^1 *if and only if* f *is an outer function of norm* 1.

Proof. First, let us show that if $||f||_1 = 1$ and f is outer, then f is extreme. Suppose that f is any L^1 function on the circle which has L^1 norm 1 and which does not vanish on a set of positive measure. Let g be a function in L^1 such that $||f \pm g||_1 \leq 1$. Of course, equality must hold, i.e.,

$$||f + g||_1 = ||f - g||_1 = ||f||_1 = 1.$$

Let $\phi = g/f$ so that ϕ is integrable with respect to the measure $d\mu = \dfrac{1}{2\pi}|f|d\theta$. Now μ is a positive measure of mass 1 and the conditions assumed on g state that

$$\int [|1 + \phi| + |1 - \phi|]d\mu = 2.$$

But

$$|1 + \phi| + |1 - \phi| \geq 2$$

and, since μ has mass 1, we must have

$$|1 + \phi| + |1 - \phi| = 2$$

almost everywhere. But this implies that ϕ is real-valued and $-1 \leq \phi \leq 1$ almost everywhere. Thus, any g in L^1 such that $||f \pm g||_1 \leq 1$ has the form $g = \phi f$, where $-1 \leq \phi \leq 1$.

Now suppose f is in H^1 and $||f||_1 = 1$. Then f does not vanish on a set of positive measure on the circle; consequently, if g is in H^1 and $||f \pm g|| \leq 1$, we have (on the circle) $g = \phi f$, where $-1 \leq \phi \leq 1$. Thus $|g| \leq |f|$ on the circle, and if f is outer this implies that

$$|g(z)| \leq |f(z)|$$

on the unit disc. Therefore, $\phi = g/f$ is a bounded analytic function with real boundary values and is constant. But then since

$$(1 - \phi)||f||_1 = (1 + \phi)||f||_1 = 1$$

we see that $\phi = 0$, that is, $g = 0$. Thus, f is extreme in the unit ball of H^1.

Now suppose f is a non-zero function in H^1 which is *not* an outer function. This means that $f = IF$, where I is a non-constant inner function and F is an outer function. Recall that this factorization of f is unique up to a constant of modulus 1, that

$$F(z) = \lambda \exp\left[\frac{1}{2\pi}\int_{-\pi}^{\pi} \frac{e^{i\theta} + z}{e^{i\theta} - z} \log |f(e^{i\theta})|d\theta\right], \quad |\lambda| = 1$$

and that $I = f/F$. Let us adjust the constant λ occurring in the outer part F so that

$$\int_{-\pi}^{\pi} |f(e^{i\theta})| \, \text{Re } I(e^{i\theta})d\theta = 0.$$

Now let

$$g = \tfrac{1}{2}(1 + I^2)F.$$

Then g is a non-zero ($I \neq$ const) function in H^1. Now, since

$$|I(e^{i\theta})| = 1 \quad \text{and} \quad 2 \operatorname{Re}(z) = \frac{1+z^2}{z} \quad \text{for } |z| = 1$$

we see that
$$g(e^{i\theta}) = f(e^{i\theta}) \operatorname{Re} I(e^{i\theta}).$$

Thus,
$$|f(e^{i\theta}) \pm g(e^{i\theta})| = |f(e^{i\theta})|[1 \pm \operatorname{Re} I(e^{i\theta})].$$

Consequently,
$$\|f \pm g\|_1 = \|f\|_1$$

because we arranged that $|f| \operatorname{Re} I$ has integral 0 on the circle. This argument shows that if $\|f\|_1 \leq 1$ and f is not outer, then f is not an extreme point of the unit ball in H^1.

We should like now to give some corollaries (also due to Rudin and de Leeuw) to this theorem and its proof, which will exhibit some rather fascinating facts about the unit ball in H^1. Probably the most remarkable is the following.

Corollary 1. *Let* f *be a function in the unit ball* Σ *of* H^1.

(i) *If* $\|f\|_1 = 1$ *and* f *is not an extreme point of* Σ *then* $f = \frac{1}{2}(f_1 + f_2)$, *where* f_1 *and* f_2 *are distinct extreme points of* Σ.

(ii) *If* $\|f\|_1 < 1$, *then* f *is a convex combination of some two extreme points of* Σ.

Proof. (i) Suppose $\|f\|_1 = 1$. We construct g as in the proof of the above theorem and put $f_1 = f + g$, $f_2 = f - g$. Then

$$\|f_1\|_1 = \|f_2\|_1 = 1 \quad \text{and} \quad f = \tfrac{1}{2}(f_1 + f_2).$$

By the theorem, we are done if we show that f_1 and f_2 are outer functions. The claim is that $f \pm tg$ are outer functions for any real t satisfying $t \geq 1$. Now

$$f \pm tg = f \pm \frac{t}{2}(1 + I^2)F$$
$$= \tfrac{1}{2}F(\pm tI^2 + 2I \pm t)$$

by the definition of g and the fact that $f = IF$. Now we choose a real number s such that $\cos s = -\frac{1}{t}$. Then

$$(\pm tI^2 + 2I \pm t) = \pm t(1 \pm e^{is}I)(1 \pm e^{-is}I)$$

so
$$f \pm tg = \pm \frac{t}{2}F(1 \pm e^{is}I)(1 \pm e^{-is}I).$$

Since the product of outer functions is outer, it suffices to show that the last two factors above are outer functions. This amounts to showing that if h is an inner function then $1 + h$ is outer. This is easy to see. For if $\epsilon > 0$, then $(1 + \epsilon) + h = h_\epsilon$ is outer, being bounded and bounded away from zero. Thus

$$h_\epsilon(z) = \exp\left[\frac{1}{2\pi}\int_{-\pi}^{\pi}\frac{e^{i\theta}+z}{e^{i\theta}-z}\log|h_\epsilon(e^{i\theta})|d\theta + i\arg h_\epsilon(0)\right].$$

As $\epsilon \to 0$, the functions $\log|h_\epsilon|$ decrease monotonically to $\log|h+1|$, so by the Lebesgue dominated convergence theorem the above formula holds with h_ϵ replaced by $(1+h)$. Therefore, $1+h$ is outer.

For part (ii) of the corollary, suppose $||f||_1 = \delta < 1$. If f is zero, the argument is trivial. If $\delta > 0$ and f is outer, it is also trivial, for f is on the line segment joining the extreme points $\frac{1}{\delta}f$ and $-\frac{1}{\delta}f$. Now suppose $0 < \delta < 1$ and f is not outer. Construct g as in the theorem and then choose numbers $t_1 > 1$ and $t_2 > 1$ such that

$$||f + t_1 g||_1 = ||f - t_2 g||_1 = 1.$$

By our argument above, $f + t_1 g$ and $f - t_2 g$ are outer functions (thus extreme points), and f lies on the segment joining these functions.

Corollary 2. *The closure of the set of extreme points of the unit ball in* H^1 *consists of all* f *in* H^1 *such that*
  (a) $||f||_1 = 1$;
  (b) f *has no zeros in the open unit disc.*

Proof. Suppose f is in H^1 and $||f_n - f||_1 \to 0$, where the f_n are outer functions of norm 1. Clearly, $||f||_1 = 1$. Since $f_n \to f$ in L^1 norm on the circle, $f_n \to f$ uniformly on compact subsets of the open unit disc. Consequently, either f has no zeros on the disc or $f \equiv 0$. The latter is absurd since $||f||_1 = 1$.

Now let f satisfy (a) and (b). Put $f_r(z) = f(rz)$ for $0 < r < 1$. Then $||f_r - f||_1 \to 0$ as $r \to 1$ and f_r is outer, being bounded away from zero. If $g_r = ||f_r||_1^{-1} f_r$, g_r is an extreme point of the unit ball of H^1 and $g_r \to f$ as $r \to 1$.

Isometries

In this section we shall describe the Banach space isometries of H^1 onto itself and of H^∞ onto itself. In either case, all isometries are induced by conformal mappings of the unit disc onto itself. One can reduce the study of isometries of H^1 to the study of isometries of H^∞. This was done by de Leeuw, Rudin, and Wermer, after de Leeuw had first found the isometries of H^1 by a lengthier method. The isometries of H^∞ were also found by Nagasawa, using a different method. Through the use of the Krein-Milman theorem, the study of isometries of H^∞ can be reduced to the study of the *algebra* automorphisms of H^∞, first characterized by Kakutani. Thus, we shall begin by describing the automorphisms of H^∞.

Theorem. *Let ϕ be an automorphism of the algebra of bounded analytic functions in the unit disc onto itself. Then there is a conformal map τ of the disc onto itself such that $(\phi f)(\lambda) = f(\tau(\lambda))$ for all bounded analytic f. Conversely, for any such τ the ϕ defined in this manner is an automorphism of the algebra of bounded analytic functions in the disc.*

Proof. If τ is a conformal map, it is clear that $f \to f(\tau)$ is an automorphism of H^∞. Suppose we are given an automorphism ϕ. Since $\phi \neq 0$, it is apparent that $\phi(1) = 1$. Let f be a non-zero function in H^∞. Then f has a zero at some point of the open unit disc if and only if there is some positive integer n such that f is not the nth power of any function in H^∞. Since ϕ is an algebra automorphism, f has an nth root if and only if $\phi(f)$ has an nth root. Therefore, ϕ maps the class of functions without zeros (in the disc) onto itself. But since $\phi(1) = 1$, it follows that ϕ preserves the range of every function; for $\phi(f - \lambda) = \phi(f) - \lambda$ has a zero if and only if $f - \lambda$ has a zero.

Let z denote the identity function: $z(\lambda) = \lambda$; and let $\tau = \phi(z)$. We shall prove that τ is a conformal map of $|\lambda| < 1$ onto itself. From our remarks above, we know immediately that τ maps $|\lambda| < 1$ onto itself. We must show that τ is one-one. For each λ with $|\lambda| < 1$, the function $z - \lambda$ is prime in the ring H^∞; i.e., if $z - \lambda = fg$ with f and g in H^∞, either f or g must be invertible in H^∞. Consequently, $\phi(z - \lambda) = \tau - \lambda$ must also be prime in H^∞. Then, obviously, $\tau - \lambda$ cannot have two distinct zeros in the open disc, else $\tau - \lambda$ would be divisible by $z - \lambda_1$ and $z - \lambda_2$. Thus τ is one-one, and is a conformal map of the open disc onto itself.

Now let ψ be the automorphism of H^∞ obtained by composing ϕ with the automorphism $f \to f(\tau^{-1})$:

$$(\psi f)(\lambda) = (\phi f)(\tau^{-1}(\lambda)).$$

Then ψ is an automorphism of H^∞ such that $\psi(z) = z$. It is clear that such a ψ is the identity automorphism. Let f be in H^∞ and $|\lambda| < 1$. Then

$$f - f(\lambda) = (z - \lambda)g, \quad g \text{ in } H^\infty.$$

Since $\psi(z - \lambda) = z - \lambda$, we have

$$\psi(f) - f(\lambda) = (z - \lambda)\psi(g),$$

so $(\psi f)(\lambda) = f(\lambda)$. Since ψ is the identity,

$$(\phi f)(\lambda) = f(\tau(\lambda))$$

for all f in H^∞.

Of course, this same result holds for the algebra A of functions continuous on the closed unit disc, analytic in the interior. The proof is the same, after one makes two elementary observations: (1) If τ is a conformal map of the disc onto itself, then τ has continuous (even analytic) boundary values. Thus, $f \to f(\tau)$ is an automorphism of A. (2) If f is in A and f has no zeros in the open disc, then f has an nth root in A for each n. Another algebra to which the theorem applies is the full algebra of analytic functions in the disc.

The proof of the preceding theorem is tied to the simple connectivity of the disc, because of the argument about roots which it contains. We might present another elementary proof which is interesting and which avoids the root argument. Suppose ϕ is an automorphism of H^∞. Let f be in H^∞ and let λ be any complex number. Then λ belongs to the closure of the range of f if and only if $(f - \lambda)$ is not invertible in H^∞. Since it is clear that $\phi(1) = 1$, we see that the ranges of f and ϕf have the same closure. Let $\tau = \phi z$. The range of τ is, then, an open set whose closure is the closed unit disc (τ is non-constant since ϕ is one-one). Thus, τ maps the open disc into the open disc. Let $|\lambda| < 1$ and f in H^∞. Then $|\tau(\lambda)| < 1$ and
$$f - f(\tau(\lambda)) = [z - \tau(\lambda)]g$$
where g is in H^∞. Consequently,
$$\phi f - f(\tau(\lambda)) = (\tau - \tau(\lambda))\phi g$$
so $\phi f - f(\tau(\lambda))$ vanishes at λ, i.e.,
$$(\phi f)(\lambda) = f(\tau(\lambda)).$$
Thus ϕ is the map $f \to f(\tau)$, and, since ϕ maps H^∞ onto H^∞, it is clear that τ is a conformal map (τ^{-1} is the function which ϕ maps onto z).

Now we turn to isometries of H^∞. The result here is that if T is a linear isometry of H^∞ onto H^∞, then $T = \alpha\phi$, where α is a constant of modulus 1 and ϕ is an algebra automorphism of H^∞, that is,
$$(Tf)(\lambda) = \alpha f(\tau(\lambda))$$
for some conformal map τ. This is a consequence of a very general result on algebras of continuous functions. We shall give the proof of the general theorem. The proof will lean heavily on the theorem of Krein and Milman which we stated at the beginning of this chapter, and, as we said, we shall not prove that theorem.

Theorem. *Let* X *be a compact Hausdorff space and let* A *be a complex linear subalgebra of* C(X), *the algebra of continuous complex-valued functions on* X. *Assume that* A *contains the constant function* 1. *Suppose* T *is a one-one linear map of* A *onto* A *which is isometric:*
$$||Tf||_\infty = ||f||_\infty.$$
Then T *has the form*
$$Tf = \alpha\phi f$$
where α *is a fixed function in* A *which is of modulus* 1, $1/\alpha$ *is in* A, *and* ϕ

is an algebra automorphism. In particular, if $T(1) = 1$, *then* T *is multiplicative.*

Proof. First, let us note two things. (1) It is no loss of generality to assume that A is uniformly closed, because any isometry of A can be uniquely extended to an isometry of the uniform completion of A. (2) It is also no loss of generality to assume that A separates the points of X. We can always arrange this by identifying those points of X which are identified by every f in A.

Now, regarding A as a Banach space with the sup norm, we let Σ denote the unit ball in the conjugate space A^*. We wish to call attention to some particular elements of Σ, the point evaluations:

$$L_x(f) = f(x), \quad x \text{ in } X.$$

The Krein-Milman theorem guarantees that Σ has extreme points. We wish to show that every extreme point L of Σ has the form $L = \lambda L_x$ where $|\lambda| = 1$ and L_x is one of the above point evaluations. Suppose L is extreme in Σ. Then $||L|| = 1$, and the Hahn-Banach theorem guarantees the existence of a bounded linear functional F on $C(X)$, which is an extension of L, and for which $||F|| = ||L|| = 1$. Let S be the set of all such extensions of L. It is clear that S is a convex and weak-star compact subset of $C(X)^*$. Choose F, an extreme point of S (using Krein-Milman again). Then F is actually extreme in the unit ball of $C(X)^*$. If $F = \frac{1}{2}(F_1 + F_2)$ with $||F_j|| \leq 1$, then

$$||F_1|| = ||F_2|| = 1$$

and if L_j is the restriction of F_j to A, each L_j is in Σ and $L = \frac{1}{2}(L_1 + L_2)$. Since L is extreme, $L_1 = L_2 = L$. Thus, each F_j is a norm preserving extension of L (i.e., is in S) and because F is extreme in S, $F_1 = F_2 = F$. The extreme points in the unit ball of $C(X)^*$ are easily identified. This unit ball is the set of all complex Baire measures on X which have total variation at most one. The extreme points are, therefore, of the form $d\mu = \lambda d\delta_x$, where $|\lambda| = 1$ and δ_x is the unit point mass at some point x in X. It follows that our extreme functional F has the form

$$F(f) = \lambda f(x)$$

and, restricting to A, the extreme functional L in Σ has the form

$$L = \lambda L_x, \quad |\lambda| = 1, \quad x \in X.$$

Since A separates the points of X, the point x which is associated with L in this way is unique; therefore, λ is unique as well.

Now let E be the set of extreme points of Σ. For every f in A

$$||f||_\infty = \sup_{L \in E} |L(f)|.$$

This we see as follows. Since each L in E is a linear functional on A of norm 1,
$$||f||_\infty \geq \sup |L(f)|, \quad L \text{ in } E.$$
Suppose $||f||_\infty = 1$. Since X is compact, f attains its maximum on X at some point x_0, say, $f(x_0) = ||f||_\infty = 1$. Let
$$M = \{L \in \Sigma; L(f) = 1\}.$$
Then M is non-empty (L_{x_0} is in M); M is convex; and M is weak-star compact. By Krein-Milman, M has an extreme point L. Such an L must actually be extreme in the unit ball Σ. For if $L = \frac{1}{2}(L_1 + L_2)$ with L_j in Σ, the facts that $||f||_\infty = 1$ and $L(f) = 1$ imply that $L_j(f) = 1$, $j = 1, 2$. Then, since L is extreme in M, we have $L_1 = L_2 = L$. Thus, the "maximum set" M for f contains an extreme point of Σ, and
$$||f||_\infty = \sup_{L \in E} |L(f)|$$
the maximum actually being attained.

At last we consider the given Banach space isometry T of A. The adjoint T^* of T, defined by
$$(T^*L)(f) = L(Tf)$$
is then an isometry of A^* onto A^*. Thus T^* must carry the unit ball Σ onto itself, and also must carry the set E of extreme points of Σ onto itself. Now let
$$B = \{x \in X; L_x \text{ is extreme in } \Sigma\}.$$
Since L_x is extreme if and only if λL_x is extreme for all λ of modulus 1, the points of E are exactly the functionals λL_x with $|\lambda| = 1$ and x in B. If x is in B, then T^*L_x must also be an extreme point of Σ. This associates with x a complex number $\alpha(x)$ of modulus 1 and a point $\tau(x)$ in B such that
$$T^*L_x = \alpha(x)L_{\tau(x)}$$
or
$$(Tf)(x) = \alpha(x)f(\tau(x)), \quad \text{all } f \text{ in } A.$$
Taking $f = 1$ in the above, we see that $\alpha(x) = (T1)(x)$ for x in B, i.e., that α is (the restriction to B of) a function in A.

Now we begin to use the ring structure of A. Let f, g be in A and let x be in B. Then
$$(T[fg])(x) = \alpha(x)(fg)(\tau(x))$$
$$= \alpha(x)f(\tau(x))g(\tau(x)).$$
If we multiply by $\alpha(x)$, we obtain
$$\alpha(x)(T[fg])(x) = (Tf)(x)(Tg)(x).$$
For any h in A we know that $||h||_\infty$ is the supremum of $|L(h)|$ for L in E, that is,

$$\|h\|_\infty = \sup_{x \in B} |h(x)|.$$

We have just seen that for any f and g in A the function
$$h = \alpha T(fg) - (Tf)(Tg)$$
vanishes on B. Consequently, we have
$$\alpha T(fg) = (Tf)(Tg)$$
for all f, g in A. If we take $f = g = T^{-1}(1)$, we see immediately that $1/\alpha$ is in A. Since both α and α^{-1} have modulus 1 on B and since each attains its maximum on B, we see that $|\alpha(x)| = 1$ on all of X. Then, if we define ϕ by
$$\phi(f) = \alpha^{-1} Tf$$
it is clear that ϕ is an algebra automorphism of A.

Corollary. *Let T be any Banach space isometry of H^∞ onto H^∞. Then T has the form*
$$(Tf)(\lambda) = \alpha f(\tau(\lambda)), \quad f \text{ in } H^\infty$$
where α is a complex constant of modulus 1 and τ is a conformal map of the open unit disc onto itself.

Proof. We shall (of course) apply the above theorem. In order to do so, we must show that H^∞ is isometrically isomorphic to an algebra of continuous functions on a compact Hausdorff space X. One way to do this is the following. For each λ, $|\lambda| < 1$, let L_λ be the complex homomorphism of H^∞ defined by $L_\lambda(f) = f(\lambda)$. Let X be the weak-star closure of the set of all such L_λ in the conjugate space $(H^\infty)^*$. Since each L_λ is in the unit ball of $(H^\infty)^*$, we see that X is a compact Hausdorff space when it is endowed with the weak-star topology. With each f in H^∞ associate the function $\hat{f}$ on X by
$$\hat{f}(L) = L(f), \quad L \text{ in } X.$$
Each $\hat{f}$ is continuous on X by definition of the weak-star topology. Because each L_λ is multiplicative on H^∞, it is clear that $f \to \hat{f}$ is a (sup-norm) isometric isomorphism of H^∞ with an algebra of continuous functions on X.

If we now apply the theorem, every isometry of H^∞ has the form $Tf = \alpha \phi f$, where ϕ is an algebra automorphism of H^∞ and α is a function in H^∞ which is of modulus 1 on the unit disc. Such an α is constant, and, as we proved, ϕ has the form
$$(\phi f)(\lambda) = f(\tau(\lambda))$$
for some conformal map τ. Hence, we are done.

Once again, we point out that the same corollary holds for the algebra A of continuous functions on the closed disc which are analytic in the interior. The proof is even shorter, since A is already an algebra of continuous functions on a compact space.

Theorem. *Every linear isometry of* H^1 *onto* H^1 *is of the form*
$$(Tf)(\lambda) = \alpha \tau'(\lambda) f(\tau(\lambda))$$
where α is a complex constant of modulus 1 and τ is a conformal map of the open unit disc onto itself. Conversely, for any such α and τ the T so defined is an isometry of H^1 onto H^1.

Proof. The converse statement is readily verified. Suppose, then, that T is a linear isometry of H^1 onto H^1. Then T preserves the unit ball in H^1 and preserves its extreme points. Consequently, T preserves the closure of the set of extreme points. But this is the set of functions of norm 1 which have no zeros in the open unit disc. We conclude that, for f in H^1, f has a zero if and only if Tf has a zero.

Now let $F = T(1)$. By our remarks above, F has no zeros in the open disc. Similarly, if f is in H^1 and λ is a complex number, $f - \lambda$ has a zero if and only if $Tf - \lambda F$ has a zero. In other words, f and Tf/F have the same range on the open disc. Define
$$Uf = \frac{Tf}{F}.$$
Then U maps the subspace H^∞ onto itself; in fact, U is isometric on H^∞ (with the sup norm), since Uf and f have the same range. Actually, U maps H^∞ onto H^∞, because if g is in H^∞ the function $f = T^{-1}(Fg)$ is in H^∞ and $Uf = g$. Furthermore, $U(1) = 1$. Hence, U is an algebra automorphism of H^∞ and has the form
$$(Uf)(\lambda) = f(\tau(\lambda))$$
for some conformal map τ. This says that
$$(Tf)(\lambda) = F(\lambda) f(\tau(\lambda))$$
for every bounded f in H^1.

From the last expression for T, we have
$$\|Tf\|_1 = \frac{1}{2\pi} \int_{-\pi}^{\pi} |F(e^{i\theta})| \, |f(\tau(e^{i\theta}))| d\theta$$
for all f in H^∞. But for any f in H^1
$$\|f\|_1 = \frac{1}{2\pi} \int_{-\pi}^{\pi} |f(e^{i\theta})| d\theta = \frac{1}{2\pi} \int_{-\pi}^{\pi} |\tau'(e^{i\theta})| \, |f(\tau(e^{i\theta}))| d\theta.$$
Since T is isometric, we have
$$\int_{-\pi}^{\pi} |F(e^{i\theta})| \, |g(e^{i\theta})| d\theta = \int_{-\pi}^{\pi} |\tau'(e^{i\theta})| \, |g(e^{i\theta})| d\theta$$
for every g in H^∞. So
$$\int |F| p \, d\theta = \int |\tau'| p \, d\theta$$

for every non-negative bounded measurable function p, because $p + \epsilon = |g|$ with $g \in H^\infty$ for any $\epsilon > 0$ [log $(p + \epsilon)$ is integrable]. Thus $|F| = |\tau'|$ a.e. on the circle. But both F and τ' are outer functions; F is outer because $F = T(1)$ and 1 is outer; τ' is outer by direct computation of the derivative of a linear fractional map. It follows that $|F(\lambda)| = |\tau'(\lambda)|$ for $|\lambda| < 1$, or that $F(\lambda) = \alpha \tau'(\lambda)$, where α is a constant of modulus 1. We now know that T has the desired form on H^∞, and, since H^∞ is dense in H^1, we are done.

To the best of our knowledge, the description of the isometries of H^p for $p \neq 1, 2, \infty$ remains unknown. In particular, one might ask whether every Banach space isometry of H^p for $p \neq 2$ is induced by a conformal map of the disc onto itself.

Projections from L^p to H^p

One of the most useful features of a Hilbert space H arises from the fact that each closed subspace S of H has a natural complementary closed subspace, $S^\perp$. The direct sum decomposition $H = S \oplus S^\perp$ associates with S a bounded projection P, which has S as range and $S^\perp$ as null space. This is all so natural in a Hilbert space that one virtually identifies P and S. In our early treatment we were (often indirectly) aided by this, in the situation where $H = L^2$ and $S = H^2$.

If X is a Banach space and S is a closed subspace of X, there is not usually any closed subspace T such that $X = S \oplus T$, let alone a "natural" T with this property. In the pure linear space sense, there are many subspaces T which are complementary to S. It is only when we require T to be closed that any difficulty arises. If $X = S \oplus T$, we have a unique projection from X onto S with null space T:

$$x = s + t$$

$$Px = s.$$

If T is closed, then P is a bounded (i.e., continuous) linear operator on X; and, although T may not be "natural," the existence of this bounded projection of X onto S can be very useful.

The situation we shall discuss is that in which the Banach space is L^p of the unit circle, and the closed subspace is H^p ($1 \leq p \leq \infty$). In this context there is certainly a natural candidate for a closed subspace complementary to H^p, namely, the space of complex conjugates of H^p functions vanishing at the origin. In other words, if f is in L^p and

$$f \sim \sum_{n=-\infty}^{\infty} c_n e^{in\theta}$$

then one might hope that $f = g + h$, where

$$g \sim \sum_{n=0}^{\infty} c_n e^{in\theta}$$

$$h \sim \sum_{n=-\infty}^{-1} c_n e^{in\theta}$$

and that this splitting of the Fourier series of f really defines two functions g and h in L^p. This is certainly the case if $p = 2$. In fact, we shall prove below the theorem of M. Riesz, which states that it is the case for $1 < p < \infty$; however, for $p = 1$ and $p = \infty$ it fails utterly. For example, it is very easy to see that

$$\sum_{n=1}^{\infty} \frac{\sin n\theta}{n}$$

is the Fourier series of a bounded function on the circle, but that

$$\sum_{n=1}^{\infty} \frac{1}{n} e^{in\theta}$$

is not the Fourier series of a bounded function, i.e., a function in H^∞. With a bit more work one can show that

$$\sum_{n=2}^{\infty} \frac{\cos n\theta}{\log n}$$

is the Fourier series of an L^1 function; but, as we proved earlier,

$$\sum_{n=2}^{\infty} \frac{1}{\log n} e^{in\theta}$$

is not the Fourier series of an H^1 function $\left(\sum \frac{1}{n \log n} = \infty \right)$.

Suppose we let $\overline{H}_0^p$ be the space of complex conjugates of H^p functions vanishing at the origin. Certainly, $\overline{H}_0^p$ is a closed subspace of L^p. As we said, we shall see that

$$L^p = H^p \oplus \overline{H}_0^p, \quad 1 < p < \infty.$$

For $p = 1$ we have

$$L^1 \neq H^1 \oplus \overline{H}_0^1.$$

Of course, the set of functions f in L^1 which are of the form $f = g + \overline{h}$ with g in H^1 and h in H_0^1 is dense in L^1, since the trigonometric polynomials are so decomposable. But the map

$$\sum_{n=-N}^{N} c_n e^{in\theta} \to \sum_{n=0}^{N} c_n e^{in\theta}$$

is not bounded in the L^1 norm. Later, we shall prove D. J. Newman's result that H^1 has no complementary closed subspace in L^1 whatsoever.

For $p = \infty$ the situation is (in a certain sense) even worse. As we noted,

$$L^\infty \neq H^\infty \oplus \overline{H}_0^\infty$$

but, furthermore, the functions of the form $g + \bar{h}$ with g and h in H^∞ are not even dense in L^∞.

Theorem. *The closed linear span of the functions in* H^∞ *and their complex conjugates is not all of* L^∞.

Proof. This proof was suggested to us by D. J. Newman. For any f in L^∞ let

$$\tilde{f}(re^{i\theta}) = \frac{1}{2\pi} \int_{-\pi}^{\pi} f(\theta - t) Q_r(t) dt$$

where Q_r is the conjugate Poisson kernel. If f is real-valued, then $\tilde{f}$ is the harmonic function vanishing at the origin which is conjugate to the harmonic extension of f. The idea of the proof is this. For any f in L^∞ which is in the closed linear span of H^∞ and $\bar{H}^\infty$, the function $\tilde{f}$ does not grow very rapidly along the positive axis. To be specific,

$$\tilde{f}(r) = o\left(\log \frac{1}{1-r}\right), \quad r \to 1^-.$$

On the other hand, if $f(\theta) = \frac{1}{2}(\pi - \theta)$ for $0 \leq \theta \leq 2\pi$ the Fourier series for f is

$$\sum_{n=1}^{\infty} \frac{\sin n\theta}{n}$$

so

$$|\tilde{f}(r)| = \log\left(\frac{1}{1-r}\right)$$

demonstrating that the little-oh relation does not hold for every f in L^∞.

All we need demonstrate is that the growth condition on $\tilde{f}$ is satisfied for functions in the closure of $H^\infty + \bar{H}^\infty$. This is clearly the case if f is in $H^\infty + \bar{H}^\infty$, since $\tilde{f}$ is then a bounded function in the disc. Since

$$Q_r(t) = \frac{2r \sin t}{1 - 2r \cos t + r^2}$$

it is clear that the functions

$$\frac{Q_r(t)}{\log(1-r)}$$

are bounded in L^1 norm as $r \to 1$. Thus, if $f_n \to f$ in the essential sup norm

$$\frac{\tilde{f}_n(r)}{\log(1-r)} \to \frac{\tilde{f}(r)}{\log(1-r)}$$

uniformly in r, and the growth condition holds for f in the closure of $H^\infty + \bar{H}^\infty$.

Let us turn now to the main positive result.

Theorem (M. Riesz). *Let* $1 < p < \infty$. *Let* f *be an* L^p *function on the unit circle with Fourier series*

Then
$$\sum_{n=-\infty}^{\infty} c_n e^{in\theta}.$$

$$\sum_{n=0}^{\infty} c_n e^{in\theta}$$

is the Fourier series of an L^p function g, and

$$\|g\|_p \leq K_p \|f\|_p$$

where K_p is a constant depending only on p, not on f. Equivalently, if $h = u + iv$ is in H^p and v vanishes at the origin,

$$\|v\|_p \leq M_p \|u\|_p$$

where M_p is a constant independent of h.

Proof. Let us first comment on the equivalence of the two assertions. It clearly suffices to prove the assertions when f and h are trigonometric polynomials. In the first case, we may also assume that f is real-valued, i.e., that $f = h + \bar{h}$, where h is a polynomial; the first statement says that the map from u to h is bounded in the L^p norm, while the second says that the map from u to v is bounded. These are clearly equivalent.

For the proof, assume that $h = u + iv$ is an analytic function with continuous boundary values and that v vanishes at the origin. We prove

$$\|v\|_p \leq M_p \|u\|_p.$$

First assume that $u > 0$ so that h maps the disc into the right half-plane. A suitable branch of h^p is then analytic with continuous boundary values. Remembering that $v(0) = 0$, we have

$$\frac{1}{2\pi} \int h^p d\theta = h^p(0)$$
$$= [h(0)]^p$$
$$= \left[\frac{1}{2\pi} \int (u + iv) d\theta\right]^p$$
$$= \left[\frac{1}{2\pi} \int u d\theta\right]^p$$
$$\leq \frac{1}{2\pi} \int u^p d\theta.$$

Now we estimate $\int |h^p - (iv)^p| d\theta$. Write

$$(u + iv)^p - (iv)^p = \int_{iv}^{u+iv} pz^{p-1} dz$$

and then

$$|h^p - (iv)^p| \leq pu|u + iv|^{p-1}$$
$$= pu(u^2 + v^2)^{(p-1)/2}.$$

Since $(A+B)^r \leq 2^r(A^r+B^r)$ for $r > 0$, we have
$$|h^p - (iv)^p| \leq pu2^{(p-1)/2}[u^{p-1} + |v|^{p-1}].$$
Accordingly,
$$\frac{1}{2\pi}\int |h^p - (iv)^p|d\theta \leq p2^{(p-1)/2}\left[\frac{1}{2\pi}\int u^p d\theta + \frac{1}{2\pi}\int u|v|^{p-1}d\theta\right]$$
$$\leq p2^{(p-1)/2}[||u||_p^p + ||u||_p||v||_p^{p-1}]$$
by applying Hölder's inequality to
$$\int u|v|^{p-1}d\theta.$$
Combining this with our previous estimate, we obtain
$$\frac{1}{2\pi}\int |(iv)^p|d\theta \leq \frac{1}{2\pi}\int |h^p - (iv)^p|d\theta + \frac{1}{2\pi}\int |h|^p d\theta$$
$$\leq p2^{(p-1)/2}[||u||_p^p + ||u||_p||v||_p^{p-1}] + ||u||_p^p.$$
Now
$$(iv)^p = |v|^p \exp[\pm\tfrac{1}{2}ip\pi]$$
so
$$\operatorname{Re}(iv)^p = |v|^p \cos(\tfrac{1}{2}p\pi).$$
But
$$\left|\frac{1}{2\pi}\int \operatorname{Re}(iv)^p d\theta\right| = \left|\operatorname{Re}\frac{1}{2\pi}\int (iv)^p d\theta\right| \leq \left|\frac{1}{2\pi}\int (iv)^p d\theta\right|$$
so we obtain
$$\left|\cos\left(\frac{p\pi}{2}\right)\right|\frac{1}{2\pi}\int |v|^p d\theta \leq \frac{1}{2\pi}\int |(iv)^p|d\theta.$$
If we combine this with our estimate for the latter integral and write
$$\lambda = \frac{||v||_p}{||u||_p}$$
we have
$$\left|\cos\left(\frac{p\pi}{2}\right)\right|\lambda^p \leq p2^{(p-1)/2}(1+\lambda^{p-1}) + 1.$$

If $p \neq 3, 5, 7, \ldots$ so that $\cos\left(\frac{p\pi}{2}\right) \neq 0$, we conclude that $||v||_p/||u||_p$ is bounded for $u > 0$. It is easy to remove the restriction $u > 0$. Given an $h = u + iv$, write $u = u^+ - u^-$, where u^+ and u^- are positive continuous functions. If we approximate u^+ and u^- uniformly by real parts of analytic functions with continuous boundary values, a simple application of the triangle inequality (plus the bound for $u > 0$) shows that $||v||_p/||u||_p$ is bounded.

This completes the proof except for the exceptional values $p = 3, 5,$

7, ... For such a p, let q be the conjugate exponent, $\dfrac{1}{p} + \dfrac{1}{q} = 1$. Then we have the theorem for q.

Note that
$$\|v\|_p = \sup \left| \frac{1}{2\pi} \int vs\,d\theta \right|$$
where s ranges over all real parts of analytic functions with continuous boundary values (or all real parts of polynomials) satisfying $\|s\|_q = 1$. For such an s let t be the harmonic conjugate vanishing at the origin. Then
$$\int vs\,d\theta = -\int ut\,d\theta$$
because $(u+iv)(s+it)$ is real at the origin. Consequently,
$$\left| \frac{1}{2\pi} \int vs\,d\theta \right| \leq \|u\|_p \|t\|_q$$
$$\leq \|u\|_p M_q \|s\|_q$$
$$= M_q \|u\|_p.$$

That completes the proof.

Theorem (D. J. Newman). *There is no closed subspace of L^1 complementary to H^1. Equivalently, there is no bounded projection from L^1 onto H^1.*

Proof (Rudin). The idea of the proof is to show that if there is any bounded projection of L^1 onto H^1 then the "natural" projection must be bounded. So assume we have a linear transformation P which associates with each f in L^1 a function Pf in H^1 in such a way that
   (i) $\|Pf\|_1 \leq K\|f\|_1$, K a positive constant;
   (ii) $Pf = f$ for each f in H^1.

Now, for f in L^1 define
$$\tilde{P}f = \frac{1}{2\pi} \int_{-\pi}^{\pi} (Pf_\theta)_{-\theta}\,d\theta.$$
Here, f_θ denotes the θ-translate (or rotation) of f:
$$f_\theta(\psi) = f(\theta + \psi).$$
For a fixed f in L^1, the map $\theta \to f_\theta$ is a continuous function from the circle into L^1, and since P is continuous, the map $\theta \to (Pf_\theta)_{-\theta}$ is continuous. This leaves no confusion as to the meaning of the integral defining $\tilde{P}f$. Now it is clear that $\tilde{P}$ is a bounded linear operator on L^1. Since H^1 is rotation invariant it is evident that $\tilde{P}$ maps L^1 into H^1; in fact, a short computation shows that
$$\tilde{P}(e^{in\theta}) = \begin{cases} e^{in\theta}, & n \geq 0 \\ 0, & n < 0. \end{cases}$$
We conclude that $\tilde{P}$ is the "natural projection" from L^1 to H^1, at least on

trigonometric polynomials. Since that projection is not bounded we have a contradiction.

The proof above works equally well for projections from L^p to H^p as long as $1 \leq p < \infty$; that is, the proof shows that if there is any bounded projection from L^p to H^p, then the natural projection is bounded. This is because rotation is continuous in the L^p-norm. The difference is that for $1 < p < \infty$ we have M. Riesz's theorem that the natural projection is bounded, whereas for $p = 1$ it is not. The proof also carries over directly to projections from the space C of continuous functions on the circle to the space A of continuous functions whose negative Fourier coefficients vanish. For, rotation $\theta \to f_\theta$ is continuous in the sup norm, if f is a continuous function. We conclude that there is no bounded projection from C onto A.

For the case of L^∞ and H^∞ the argument is not quite so simple, since rotation is not continuous in the sup norm for a fixed L^∞ function; however, this only forces a minor modification in the proof.

Theorem. *There is no bounded projection of* L^∞ *onto* H^∞.

Proof. Suppose P is such a bounded projection. Let us employ the notation

$$[f, g] = \frac{1}{2\pi} \int fg d\theta.$$

If f is in L^∞ we define a function $\tilde{P}f$ in L^∞ by

$$(\tilde{P}f, g) = \frac{1}{2\pi} \int_{-\pi}^{\pi} [Pf_\theta, g_\theta] d\theta, \quad \text{all } g \text{ in } L^1.$$

Since P is bounded, for a fixed f the integral on the right defines a bounded linear functional on L^1, and this linear functional is therefore "inner product with" some function in L^∞. The latter function we call $\tilde{P}f$. Now one can proceed as before to verify that $\tilde{P}$ is the natural projection of L^∞ to H^∞ and is also bounded—a contradiction.

The proof of this last result was first shown to us by R. Arens, who with P. C. Curtis had independently found Rudin's proof about projections from L^p to H^p. The informed reader will note that the last theorem implies that there is no bounded projection from L^1 to H^1. If there were a closed subspace of L^1 complementary to H^1, the conjugate space L^∞ would be the direct sum of the annihilator of that closed space and the annihilator of H^1. Thus there would be a closed subspace of L^∞ complementary to the space of H^∞ functions vanishing at the origin.

NOTES

Kakutani's theorem on automorphisms is in [50]. The characterization of the extreme points in the unit ball is due to Arens, Buck, Carleson, and Hoffman, during the 1957 Conference on Analytic Functions at the Institute for Advanced Study. The extreme points of the unit ball in H^1 were found by deLeeuw and Rudin [23]. This paper also contains the two corollaries to the characterization, as well as a discussion of some other extremum problems in H^1. The Krein-Milman theorem is in [52]. For a proof, one may consult the book of Dunford and Schwartz [25]. The theorem which we proved here concerning isometries of function algebras first came to my attention through the paper of Arens and Singer [5]; however, it has been discovered independently by a great many people, and I know not to whom original credit should go. The isometries of H^1 were first found by deLeeuw [22]. The proof presented here is due to deLeeuw, Rudin, and Wermer [24], all of their work being carried on independent of Nagasawa [63], who also described the isometries of H^∞. M. Riesz's proof of the boundedness of the natural projection from L^p to H^p $(1 < p < \infty)$ can be found in Zygmund [98], along with various other proofs of the theorem. See Bochner [14] for another proof. As we gave it, Riesz's proof carries over directly to the context of Dirichlet algebras, as was pointed out to me by J. Wermer. Let A be a Dirichlet algebra on the compact Hausdorff space X, and let m be a positive measure on X which is multiplicative on A. Assume each real-valued f in A is constant. If u is the real part of a function in A, there is a unique v such that $\int v\,dm = 0$ and $u + iv$ is in A. The theorem, then, says that the map from u to v is bounded in the $L^p(dm)$ norm, for $1 < p < \infty$. D. J. Newman's paper on the impossibility of projecting L^1 onto H^1 is [66]. Rudin [77] first found the proof given here, and has generalized the result somewhat.

EXERCISES

1. Let f be in A, i.e., f is continuous on the closed unit disc analytic in the interior. If f has no zeros in $|z| < 1$, prove that f has an nth root in A, for $n = 1, 2, 3, \ldots$.

2. Prove that the unit ball in L^1 has no extreme points.

3. If $1 < p < \infty$, prove that every function of norm 1 is an extreme point of the unit ball in L^p.

4. Prove that every element of norm 1 in a Hilbert space H is an extreme point of the unit ball in H.

5. (deLeeuw-Rudin) Regard H^1 as a subspace of the conjugate space of the continuous functions on the circle. Prove that the weak-star closure of the set of extreme points of the unit ball in H^1 consists of all f in H^1 such that $||f||_1 \leq 1$ and f has no zeros in the open unit disc, together with the zero function.

6. Let K be a closed subset of the unit circle, and let B_K be the subspace of functions in H^∞ which are continuous at each point of K. Find all (sup-norm) isometries of B_K onto B_K.

7. Consider the second proof we gave that every automorphism of H^∞ is induced by a conformal map. Use it to prove that if D is any bounded region in the plane which is the interior of its closure, every automorphism of the algebra of bounded analytic functions on D is induced by a conformal map D onto D.

8. Let D and R be bounded regions in the plane. Let ϕ be a homomorphism of the algebra of bounded analytic functions on D onto the algebra of bounded analytic functions on R, $\phi \neq 0$.

(a) Show that $\phi(1)$ is either 1 or 0 on each component of R; and if R_0 is the union of those components on which $\phi(1) = 1$, then ϕ is really a homomorphism into the bounded analytic functions on R_0 (so that we may as well assume that $\phi(1) = 1$).

(b) If $\phi(1) = 1$, and $\tau = \phi(z)$, then $\tau(R)$ is contained in $\overline{D}$ (the closure of D).

(c) If the interior of $\overline{D}$ is D, $\phi(1) = 1$, and τ is non-constant, then

$$(\phi f)(\lambda) = f(\tau(\lambda)).$$

(d) If ϕ is one-one and onto and $int\, \overline{D} = D$, then

$$(\phi f)(\lambda) = f(\tau(\lambda))$$

where τ is a conformal map of R onto D.

9. [98; page 253] Prove that $\sum_{n=2}^{\infty} \dfrac{\cos n\theta}{\log n}$ is the Fourier series of an integrable function, but $\sum_{n=2}^{\infty} \dfrac{e^{in\theta}}{\log n}$ is not in H^1.

10. [98; page 253] Prove that $\sum_{n=1}^{\infty} \dfrac{\sin n\theta}{n}$ is the Fourier series of a bounded function, but that $\sum_{n=1}^{\infty} \dfrac{1}{n} e^{in\theta}$ is not in H^∞.

11. [98; page 253] Prove that $\sum_{n=1}^{\infty} \dfrac{\sin n\theta}{n \log n}$ converges uniformly to a continuous function on the unit circle, but that $\sum_{n=1}^{\infty} \dfrac{1}{n} e^{in\theta}$ is not the Fourier series of a function in A (i.e., of a continuous function).

CHAPTER 10

H^∞ AS A BANACH ALGEBRA

In this chapter we deal with H^∞, the algebra of bounded analytic functions in the unit disc. In Chapter 9 we studied this algebra to a limited extent; we classified its automorphisms and from this classified its Banach space isometries. Here we shall concentrate on the space of maximal ideals of H^∞, relate what is known to date about its structure, and add a few new pieces of information. First, let us clarify the term "maximal ideal space."

We consider a commutative Banach algebra B (with identity). As we proved in Chapter 6, there is a one-one correspondence

$$\phi \leftrightarrow M$$

between homomorphisms ϕ of B onto the algebra of complex numbers and maximal ideals M in the algebra B. The correspondence is defined by $M = \text{kernel } (\phi)$. We also showed that each maximal ideal M is closed and consequently that each complex homomorphism ϕ is continuous:

$$|\phi(x)| \leq \|x\|.$$

Therefore, the complex homomorphisms of B are the *bounded* linear functionals on B which are multiplicative:

$$\phi(xy) = \phi(x)\phi(y).$$

The bound (or norm) of such a ϕ is precisely 1, since $\phi(1) = 1$. Suppose we let $\mathfrak{M}(B)$ denote the set of complex homomorphisms of B. Then $\mathfrak{M}(B)$ is a subset of the conjugate space B^*, and in fact is contained in the unit sphere of B^*. Also, $\mathfrak{M}(B)$ is closed in the weak-star topology on B^*. Let L be a bounded linear functional on B which is in the weak-star closure of $\mathfrak{M}(B)$. If x and y are elements of B and $\epsilon > 0$, there is an element ϕ in $\mathfrak{M}(B)$ such that

$$|L(x) - \phi(x)| < \epsilon$$
$$|L(y) - \phi(y)| < \epsilon$$
$$|L(xy) - \phi(xy)| < \epsilon.$$

Since $\phi(xy) = \phi(x)\phi(y)$, it is clear that

$$|L(xy) - L(x)L(y)| < \epsilon(1 + ||x|| + ||y||).$$

Thus L is multiplicative; i.e., L is in $\mathfrak{M}(B)$. Since the unit ball in B^* is weak-star compact, it follows that $\mathfrak{M}(B)$ is a compact Hausdorff space when it is equipped with the weak-star topology. This compact space is called the **space of complex homomorphisms** of B, or (because of the correspondence between homomorphisms and maximal ideals) it is often called the **space of maximal ideals** of B. We shall use the latter term; however, we shall continue to think of the elements of $\mathfrak{M}(B)$ as homomorphisms.

With each element x in B we associate a complex-valued function $\hat{x}$ on $\mathfrak{M}(B)$ by

$$\hat{x}(\phi) = \phi(x).$$

Each $\hat{x}$ is a continuous function on $\mathfrak{M}(B)$; indeed, by definition, the weak-star topology is the weakest (smallest) topology on $\mathfrak{M}(B)$ which makes each $\hat{x}$ continuous. If $\hat{B}$ denotes the set of all $\hat{x}$, then we have constructed a representation (algebra homomorphism)

$$x \to \hat{x}$$

from B onto $\hat{B}$, an algebra of continuous complex-valued functions on $\mathfrak{M}(B)$. This is usually called the Gelfand representation. In general, this representation is not faithful; that is, we may have $\hat{x} = 0$ but $x \neq 0$. It will, however, be an isomorphism in the cases we consider. It is always true that the representation is norm decreasing:

$$||\hat{x}||_\infty \leq ||x||.$$

This is merely a repetition of the fact that each complex homomorphism of B is bounded by 1.

One may reasonably ask what makes the representation $x \to \hat{x}$ so interesting. In the first place, perhaps the simplest commutative Banach algebras are the algebras $C(X)$ of all continuous complex-valued functions on a compact Hausdorff space X; evidently, then, it is worthwhile to represent the general B by (a subalgebra of) such an algebra. Among such representations, the Gelfand representation is fundamental because of its intimate relation to the algebraic structure of B. For example, the statement that an element x of B is invertible translates into the statement that the function $\hat{x}$ has no zeros on the space $\mathfrak{M}(B)$.

Maximal Ideals in H^∞

If we use pointwise addition and multiplication, together with the sup norm

$$||f|| = \sup_{|z|<1} |f(z)|$$

H^∞ is a commutative Banach algebra with identity. Consequently, our comments above apply to H^∞, giving us the representation $f \to \hat{f}$ of H^∞ by an algebra of continuous functions on the compact maximal ideal space $\mathfrak{M}(H^\infty)$. We wish to explore some of the structure of this space. What are the maximal ideals (complex homomorphisms) of H^∞? We shall not answer this question; indeed, it seems clear that no concrete answer could ever be given. But we shall discover enough about the structure of $\mathfrak{M}(H^\infty)$ to justify our efforts.

The only obvious complex homomorphisms of H^∞ are the point evaluations

$$\phi_\lambda(f) = f(\lambda)$$

where λ is a point in the open unit disc. It is evident that there are others. For instance, if I denotes the set of functions f in H^∞ such that $f(\lambda)$ tends to zero as λ approaches 1 along the positive axis, it is clear that I is a proper ideal in H^∞. Accordingly, I is contained in a maximal ideal; i.e., there is a complex homomorphism ϕ of H^∞ such that $\phi(f) = 0$ for all f in I. But ϕ is not one of the point evaluations ϕ_λ, since there is no λ at which every f in I vanishes. Needless to say, there are many such "new" homomorphisms; the number is impressive.

Note that the point evaluations ϕ_λ show that the Gelfand representation $f \to \hat{f}$ is one-one; if $\hat{f} = 0$, then $\hat{f}(\phi_\lambda) = f(\lambda) = 0$ for each λ in the open disc. Furthermore, we can see that the representation is isometric:

$$||f|| = ||\hat{f}||_\infty = \sup_{\phi \in \mathfrak{M}(H^\infty)} |\hat{f}(\phi)|.$$

We know that $||\hat{f}||_\infty \leq ||f||$, and the ϕ_λ tell us that

$$||\hat{f}||_\infty \geq \sup_{|\lambda|<1} |\hat{f}(\phi_\lambda)| = ||f||.$$

Thus H^∞ is isometrically isomorphic to $\hat{H}^\infty$, a uniformly closed subalgebra of the continuous complex-valued functions on the maximal ideal space $\mathfrak{M}(H^\infty)$.

There is a natural continuous mapping of $\mathfrak{M}(H^\infty)$ into the closed unit disc in the plane. If z denotes the coordinate (or identity) function on the disc, the mapping we have in mind sends the homomorphism ϕ into its value on the function z. In other words, the mapping is $\hat{z}$. To avoid confusion, let us use the symbol π for $\hat{z}$:

$$\pi(\phi) = \phi(z), \qquad \phi \in \mathfrak{M}(H^\infty).$$

Theorem. *The mapping π is a continuous map of $\mathfrak{M}(H^\infty)$ onto the closed unit disc in the plane. Over the open unit disc* D, *π is one-one, and π^{-1} maps* D *homeomorphically onto an open subset Δ of $\mathfrak{M}(H^\infty)$.*

Proof. By its very definition as $\hat{z}$, π is a continuous map from $\mathfrak{M}(H^\infty)$ into the closed unit disc. Each point λ in the open disc is in the range of π, since $\pi(\phi_\lambda) = \lambda$. Since $\mathfrak{M}(H^\infty)$ is compact, so is the range of π; therefore, this range must be the entire closed unit disc. Suppose $|\lambda| < 1$ and $\pi(\phi) = \lambda$. If f vanishes at λ, then $f = (z - \lambda)g$, and so

$$\phi(f) = \phi(z - \lambda)\phi(g) = 0 \cdot \phi(g) = 0.$$

Since $\phi(f) = 0$ for every f which vanishes at λ, it follows that ϕ is evaluation at λ. This shows that π is one-one over D. If we let $\Delta = \pi^{-1}(D)$, then π maps Δ homeomorphically onto D, since the topology of Δ is the weak topology defined by the functions $\hat{f}$, and the topology of D is the weak topology defined by the functions f in H^∞.

It is convenient to picture π as a projection of $\mathfrak{M}(H^\infty)$ onto the closed disc. As we saw, π is one-one over D, so the open unit disc is homeomorphically embedded in $\mathfrak{M}$ by $\lambda \to \phi_\lambda$. The remainder of $\mathfrak{M}$ is mapped by π onto the unit circle. If $|\alpha| = 1$, we shall call $\pi^{-1}(\alpha)$ the **fiber of $\mathfrak{M}$ over α** and denote this fiber by $\mathfrak{M}_\alpha$:

$$\mathfrak{M}_\alpha = \pi^{-1}(\alpha) = \{\phi \in \mathfrak{M}; \ \phi(z) = \alpha\}.$$

The fiber $\mathfrak{M}_\alpha$ is a closed subset of $\mathfrak{M}$ and consists of the homomorphisms of H^∞ which resemble 'evaluation at α.' Let us make this precise.

Theorem. *Let* f *be a function in* H^∞ *and let* α *be a point of the unit circle. Let* $\{\lambda_n\}$ *be a sequence of points in the open unit disc which converges to* α, *and suppose that the limit* $\zeta = \lim f(\lambda_n)$ *exists. Then there is a complex homomorphism* ϕ *in the fiber* $\mathfrak{M}_\alpha$ *such that* $\phi(\mathrm{f}) = \zeta$.

Proof. Let J be the collection of functions g in H^∞ for which $\lim g(\lambda_n) = 0$. Then J is an ideal in H^∞ and is contained in a maximal ideal, that is, there is a complex homomorphism ϕ of H^∞ such that $\phi(g) = 0$ for every g in J. The functions $(z - \alpha)$ and $(f - \zeta)$ are both in J. Thus $\phi(z) = \alpha$ and $\phi(f) = \zeta$.

Theorem. *The function* $\hat{\mathrm{f}}$ *is constant on the fiber* $\mathfrak{M}_\alpha$ *if and only if* f *is continuously extendable to* $D \cup \{\alpha\}$.

Proof. If f is continuously extendable to $D \cup \{\alpha\}$, there is a complex number ζ such that $\lim f(\lambda_n) = \zeta$ for every sequence of points λ_n in D which converges to α. We wish to show that $\hat{f}$ has the constant value ζ on the fiber $\mathfrak{M}_\alpha$. Assume $\zeta = 0$. Let $h(\lambda) = \frac{1}{2}(1 + \bar{\alpha}\lambda)$ so that $h(\alpha) = 1$ and $|h| < 1$ elsewhere on the closed unit disc. Since f is continuous at α with the value 0, one can readily check that $(1 - h^n)f$ converges uniformly to f as $n \to \infty$. If ϕ is a complex homomorphism of H^∞ which lies in the fiber $\mathfrak{M}_\alpha$, then $\phi(h) = 1$. Consequently, $\phi[(1 - h^n)f] = 0$, and, since ϕ is continuous, $\phi(f) = 0$. Thus $\hat{f}$ is identically 0 on $\mathfrak{M}_\alpha$.

If $\hat{f}$ is constant on $\mathfrak{M}_\alpha$, the last theorem shows immediately that f is continuously extendable to $D \cup \{\alpha\}$.

Theorem. *Let* f *be a function in* H^∞ *and* α *a point of the unit circle. Suppose there is a complex homomorphism* ϕ *in the fiber* $\mathfrak{M}_\alpha$ *such that* $\phi(\mathrm{f}) = 0$.

Then there is a sequence of points λ_n in D such that $\lim \lambda_n = \alpha$ and $\lim f(\lambda_n) = 0$.

Proof. If we cannot find such a sequence of points, there is an open disc N centered at α such that
$$|f(\lambda)| \geq \delta > 0, \qquad \lambda \in D \cap N.$$
Write $f = BSF$, where B is a Blaschke product, S is a singular function, and F is an outer function. Since f is bounded away from zero in a neighborhood of α, the functions B and S are analytic on that part of the unit circle C which lies in N. (The zeros of B never enter N, and the singular measure determining S must be supported on C-N.) By absorbing a constant of modulus 1 into the inner function BS, we may assume that
$$F(\lambda) = \exp\left[\frac{1}{2\pi}\int_{-\pi}^{\pi} \frac{e^{i\theta} + \lambda}{e^{i\theta} - \lambda}\log|f(e^{i\theta})|d\theta\right].$$
Now, since $|f| \geq \delta > 0$ on $D \cap N$, the function $\log|f(e^{i\theta})|$ is bounded on $N \cap C$. Thus, if we define
$$h(\lambda) = \exp\left[\frac{1}{2\pi}\int_{N \cap C} \frac{e^{i\theta} + \lambda}{e^{i\theta} - \lambda}(-\log|f(e^{i\theta})|)d\theta\right]$$
h will be a bounded analytic function. Furthermore,
$$f(\lambda)h(\lambda) = B(\lambda)S(\lambda)\exp\left[\frac{1}{2\pi}\int_{-\pi}^{\pi}\frac{e^{i\theta} + \lambda}{e^{i\theta} - \lambda}k(\theta)d\theta\right]$$
where k is integrable and vanishes identically on $N \cap C$. It follows that fh is analytically continuable across that part of the unit circle which lies in N, and also that $|fh| = 1$ on $N \cap C$. By the last theorem, the representing function $(fh)\hat{\,}$ is constant on the fiber $\mathfrak{M}_\alpha$, and that constant has modulus 1. Therefore, if ϕ is in $\mathfrak{M}_\alpha$,
$$|\phi(f)||\phi(h)| = 1$$
so $\phi(f) \neq 0$. This contradicts the hypothesis that $\phi(f) = 0$ for some ϕ in $\mathfrak{M}_\alpha$.

Corollary. *If* f *is a function in* H^∞ *and* α *is a point of the unit circle, then the range of* $\hat{f}$ *on the fiber* $\mathfrak{M}_\alpha$ *consists of all complex numbers* ζ *for which there is a sequence of points* λ_n *in* D *with* $\lim \lambda_n = \alpha$ *and* $\lim f(\lambda_n) = \zeta$.

This corollary makes precise the rough statement that each homomorphism ϕ in the fiber $\mathfrak{M}_\alpha$ is akin to "evaluation at α." It also shows us that each $\mathfrak{M}_\alpha$ contains myriads of complex homomorphisms of H^∞.

Topological Structure of $\mathfrak{M}(H^\infty)$

We shall now consider various topological questions about the maximal ideal space of H^∞. The point evaluations ϕ_λ embed the open unit disc as an open subset Δ of $\mathfrak{M}$. The remaining homomorphisms lie in the fibers $\mathfrak{M}_\alpha$, and, as the last corollary shows, are in some sense limits of the points of Δ. The question naturally arises: are these homomorphisms actually limits of the ϕ_λ in the topology of $\mathfrak{M}$? In other words, is the open disc Δ dense in $\mathfrak{M}$? This question remains unanswered. Although it seems to be quite abstract, it is easily translated into a very concrete question about bounded analytic functions. [See NOTES]

Theorem. *A necessary and sufficient condition that the open disc Δ should be dense in $\mathfrak{M}(H^\infty)$ is the following. If $f_1, \ldots, f_n$ are bounded analytic functions in the open unit disc such that*

$$|f_1(\lambda)| + \cdots + |f_n(\lambda)| \geq \delta > 0, \qquad |\lambda| < 1$$

then there exist bounded analytic functions $g_1, \ldots, g_n$ such that $f_1 g_1 + \cdots + f_n g_n = 1$.

Proof. Suppose that there is a complex homomorphism ϕ_0 of H^∞ which is not in the closure of Δ. By definition of the topology on $\mathfrak{M}$ this means that there exist functions $f_1, \ldots, f_n$ in H^∞ and a positive δ such that $\phi_0(f_j) = 0, j = 1, \ldots, n$, but the open set

$$\{\phi \in \mathfrak{M}; \ |\phi(f_j)| < \delta, \ j = 1, \ldots, n\}$$

does not intersect Δ. In particular,

$$|f_1| + \cdots + |f_n| \geq \delta$$

on the open unit disc, but $f_1, \ldots, f_n$ lie in a proper ideal of H^∞, namely, the kernel of ϕ_0. The statement that $f_1, \ldots, f_n$ lie in a proper ideal is equivalent to the statement that 1 is not in the ideal they generate, i.e., there do *not* exist $g_1, \ldots, g_n$ in H^∞ such that

$$f_1 g_1 + \cdots + f_n g_n = 1.$$

One can easily see that this argument is reversible.

We might make some comments about the status of this problem of the density of Δ in $\mathfrak{M}$. If

$$|f_1| + \cdots + |f_n| \geq \delta > 0$$

then it is relatively easy to find analytic functions $g_1, \ldots, g_n$ such that

$$f_1 g_1 + \cdots + f_n g_n = 1.$$

This requires only that $f_1, \ldots, f_n$ have no common zero in $|\lambda| < 1$. We indicated in the exercises for Chapter 6 one way to find such $g_1, \ldots, g_n$. The work comes in getting $g_1, \ldots, g_n$ to be bounded when the sum of the moduli of the f_j is bounded away from zero. In certain special cases the problem has been solved. For instance, we solved a special case in the last theorem of the preceding section. The proof which we gave of the fact that, if $\hat{f}$ vanishes on $\mathfrak{M}_\alpha$, then f tends to zero along some sequence approaching α is just the proof that if

$$|z - \alpha| + |f| \geq \delta > 0$$

then 1 is in the ideal generated by $(z - \alpha)$ and f. More difficult special cases have been solved by Newman and Carleson in their work on interpolation problems in H^∞. The set $\mathfrak{M} - \bar{\Delta}$ has been called the "corona."

Although we cannot resolve the question of whether $\mathfrak{M}-\Delta$ has an interior, we can infer a few facts about the topology of $\mathfrak{M}-\Delta$. We have the decomposition

$$\mathfrak{M}-\Delta = \bigcup_{|\alpha|=1} \mathfrak{M}_\alpha.$$

It is easy to see that the various fibers $\mathfrak{M}_\alpha$ are homeomorphic. The algebra H^∞ is rotation invariant, and each rotation induces a homeomorphism of $\mathfrak{M}$ which maps Δ onto Δ and "rotates" the fibers $\mathfrak{M}_\alpha$. More generally, let ψ be any algebra automorphism of H^∞. Then the adjoint mapping ψ^* defined by

$$(\psi^*\phi)(f) = \phi(\psi f)$$

maps $\mathfrak{M}$ homeomorphically onto $\mathfrak{M}$. We proved in the last chapter that ψ has the form

$$(\psi f)(\lambda) = f(\tau(\lambda))$$

where τ is a conformal (linear fractional) map of the disc onto itself. Thus

$$\psi^*\phi_\lambda = \phi_{\tau(\lambda)}$$

so ψ^* maps Δ onto Δ. If ϕ is in the fiber $\mathfrak{M}_\alpha$, then, clearly, $\psi^*\phi$ is in the fiber $\mathfrak{M}_{\tau(\alpha)}$, since

$$(\psi^*\phi)(z) = \phi(\psi z) = \phi(\tau) = \tau(\alpha).$$

Here we have used the fact that τ is continuous on the unit circle and the consequent fact that $\phi(\tau) = \tau(\alpha)$. Similarly, one sees that ψ^* maps $\mathfrak{M}_\alpha$ homeomorphically onto $\mathfrak{M}_{\tau(\alpha)}$.

Since the various fibers $\mathfrak{M}_\alpha$ are homeomorphic, one might guess that the decomposition

$$\mathfrak{M}-\Delta = \bigcup_{|\alpha|=1} \mathfrak{M}_\alpha$$

is a product decomposition, that is, that $\mathfrak{M}-\Delta$ is naturally homeomorphic to the topological product of the unit circle and one of the fibers $\mathfrak{M}_\alpha$. We can fix a fiber, say $\mathfrak{M}_1$, and identify $\mathfrak{M}-\Delta$ with the point set $C \times \mathfrak{M}_1$ by using rotations. Let R_α be the rotation $\lambda \to \alpha\lambda$ so that R_α^* is a homeomorphism of $\mathfrak{M}$. Then

$$\phi \leftrightarrow (\alpha, R_\alpha^*\phi)$$

is a one-one correspondence between $\mathfrak{M}$ and $C \times \mathfrak{M}_1$, but it is not a homeomorphism, because the map from α to $R_\alpha^*\phi$ is badly discontinuous. In other words, if we allow the circle group to operate on $\mathfrak{M}$ through the homeomorphisms R_α^*, it does not operate continuously, but only as a discrete group. For if we fix ϕ in $\mathfrak{M}-\Delta$, its orbit

$$\{R_\alpha^* \phi;\ \alpha \in C\}$$

under the rotation group is not a closed subset of $\mathfrak{M}$.

Theorem. *If $\{\phi_n\}$ is any sequence of points of $\mathfrak{M}$-Δ which converges, then all but a finite number of the ϕ_n lie in the same fiber $\mathfrak{M}_\alpha$. Consequently, if S is any function from the unit circle into $\mathfrak{M}$-Δ such that $\pi \circ$ S is the identity, the range of S cannot be closed; hence S cannot be continuous.*

Proof. Let $\{\phi_n\}$ be a sequence of points of $\mathfrak{M}$-Δ with ϕ_n in $\mathfrak{M}_{\alpha_n}$, i.e., $\pi(\phi_n) = \alpha_n$. If there is an infinite number of the α_n, we may as well assume that the α_n are all distinct. Choose an open arc I_n about α_n in such a way that the I_n are pairwise disjoint. Let

$$u(\alpha) = \begin{cases} (-1)^n, & \text{if } \alpha \in I_n \\ 0, & \text{otherwise.} \end{cases}$$

Define

$$f(\lambda) = \exp\left[\frac{1}{2\pi}\int_{-\pi}^{\pi} \frac{e^{i\theta} + \lambda}{e^{i\theta} - \lambda} u(e^{i\theta}) d\theta\right]$$

so that f is in H^∞ and $|f| = e^u$ on the circle C. By definition of u, we see that f is continuous on each of the arcs I_n and hence that

$$|\phi_n(f)| = \begin{cases} e, & n \text{ even} \\ \dfrac{1}{e}, & n \text{ odd.} \end{cases}$$

Certainly, then, the sequence $\{\phi_n\}$ does not converge.

The second statement of the theorem, concerning sections of the projection map π, is evident from the result about sequences in $\mathfrak{M}$-Δ.

The last statement of the theorem says that the "projection" π from $\mathfrak{M}$-Δ onto C has no continuous cross section. It certainly shows that the decomposition

$$\mathfrak{M}\text{-}\Delta = \bigcup_{|\alpha|=1} \mathfrak{M}_\alpha$$

is not a product decomposition (in the obvious way). The peculiar topological nature of this decomposition is further underscored by the following.

Theorem. *Let W_+ be the union of the fibers $\mathfrak{M}_\alpha$ for $Im\,(\alpha) > 0$ and let W_- be the union of the $\mathfrak{M}_\alpha$ for $Im\,(\alpha) < 0$. Each point in the fiber $\mathfrak{M}_1$ is in the closure of W_+, or in the closure of W_-, or in neither; these are three mutually disjoint possibilities, each of which occurs.*

Proof. Let u be the harmonic function in the unit disc which has the boundary values

$$u(\alpha) = \begin{cases} 0, & Im\,(\alpha) > 0 \\ 1, & Im\,(\alpha) < 0. \end{cases}$$

Let $f = e^{u+iv}$, where v is a harmonic conjugate of u. Then f is in H^∞ and $|f| = e^u$. Note that f is continuous at all points of the unit circle except ± 1. Therefore, the representing function $\hat{f}$ has the constant modulus 1 on W_+ and the constant modulus e on W_-. It is clear that W_+ and W_- have disjoint closures, e.g., no point in $\mathfrak{M}_1$ can lie in both of these closures, $\overline{W}_+$ and $\overline{W}_-$.

Since the projections $\pi(\overline{W}_+)$ and $\pi(\overline{W}_-)$ are compact, we see that

$$\pi(\overline{W}_+) = \{\alpha \in C;\ \operatorname{Im}(\alpha) \geq 0\}$$
$$\pi(\overline{W}_-) = \{\alpha \in C;\ \operatorname{Im}(\alpha) \leq 0\}.$$

Obviously, then, there is a point in the fiber $\mathfrak{M}_1$ which lies in $\overline{W}_+$ and also a point which lies in $\overline{W}_-$. But there are also points in $\mathfrak{M}_1$ which lie neither in $\overline{W}_+$ nor in $\overline{W}_-$. Consider the function f defined in the first part of the proof. It is evident that $|f| = e^u = \sqrt{e}$ everywhere on the unit interval. This implies the existence of a homomorphism ϕ in $\mathfrak{M}_1$ for which $|\phi(f)| = \sqrt{e}$. No such ϕ is in $\overline{W}_+$ or $\overline{W}_-$. That completes the proof.

An intuitive description of this theorem illuminates the strange character of the topology of $\mathfrak{M}\text{-}\Delta$. If we consider any fiber $\mathfrak{M}_\alpha$, the points ϕ of $\mathfrak{M}_\alpha$ fall into exactly one of these categories: ϕ can be approached from points in the fibers $\mathfrak{M}_\beta$ with β a little to one side of α; or ϕ can be approached from points in the $\mathfrak{M}_\beta$ with β a little to the other side of α; or ϕ cannot be approached by any points of $\mathfrak{M}\text{-}\Delta$ except those in the fiber $\mathfrak{M}_\alpha$. The existence of ϕ's of the third type simply shows that the fiber $\mathfrak{M}_\alpha$ has a non-empty interior *in the space* $\mathfrak{M}\text{-}\Delta$. Whether $\mathfrak{M}_\alpha$ has any interior in the maximal ideal space $\mathfrak{M}$ is unknown, since it is equivalent to whether Δ is dense in $\mathfrak{M}$, i.e., whether $\mathfrak{M}\text{-}\Delta$ has an interior.

Later, we shall establish a few more topological results about $\mathfrak{M}(H^\infty)$, chiefly that $\mathfrak{M}\text{-}\Delta$ is connected and that each of the fibers $\mathfrak{M}_\alpha$ is connected.

Discs in Fibers

In this section we shall prove the existence of a homeomorphic and analytic embedding of the open unit disc into each of the fibers $\mathfrak{M}_\alpha$. Specifically, we find a homeomorphism ψ of the open unit disc D into $\mathfrak{M}_\alpha$ such that

(i) for each f in H^∞ the composition $\hat{f} \circ \psi$ is analytic on D.
(ii) $f \to \hat{f} \circ \psi$ is an algebra homomorphism of H^∞ onto H^∞.

Let L be the linear fractional map

$$L(\lambda) = \frac{\lambda + i(\lambda - 1)}{1 + i(\lambda - 1)}.$$

Of course, L is a conformal map of the closed disc onto itself, and $\lambda = 1$

is the unique fixed point of L in the closed disc $D \cup C$. Let $L^{(n)}$ denote the nth composition of L with itself. Then

$$L^{(n)}(\lambda) = \frac{\lambda + ni(\lambda - 1)}{1 + ni(\lambda - 1)}, \qquad n = 0, \pm 1, \pm 2, \ldots.$$

Now we define a sequence of mappings from D into the maximal ideal space $\mathfrak{M}$ by

$$\psi_n(\lambda) = \pi^{-1}[L^{(2n)}(\lambda)], \qquad n = 1, 2, 3, \ldots.$$

That is, $\psi_n(\lambda)$ is the complex homomorphism of H^∞ which evaluates each f at the point $L^{(2n)}(\lambda)$. Each ψ_n is an analytic map from D into $\mathfrak{M}$; i.e., $\hat{f} \circ \psi_n$ is an analytic function on D for any f in H^∞:

$$(\hat{f} \circ \psi_n)(\lambda) = f(L^{(2n)}(\lambda)).$$

In the space of maps from D to $\mathfrak{M}$, the sequence $\{\psi_n\}$ has a cluster point ψ, because $\mathfrak{M}$ is compact. Now ψ will also be an analytic map from D into $\mathfrak{M}$, since for each f in H^∞ the sequence $\hat{f} \circ \psi_n$ is uniformly bounded and consequently uniformly equicontinuous on each compact subset of D.

Now we claim that

(i) ψ maps D into the fiber $\mathfrak{M}_1$;
(ii) ψ is a homeomorphism;
(iii) for any f in H^∞ there is a g in H^∞ such that $\hat{g} \circ \psi = f$.

Statement (i) is easily proved. We have

$$\lim_{n \to \infty} L^{(n)}(\lambda) = \lim_{n \to \infty} \frac{\lambda + ni(\lambda - 1)}{1 + ni(\lambda - 1)} = 1$$

for each λ in D. Consequently, $\pi(\psi(\lambda)) = 1$ for each λ, since $\pi(\psi(\lambda))$ is a cluster point of $L^{(2n)}(\lambda)$.

To prove statement (ii) we argue as follows. Define

$$h(\lambda) = \lambda \prod_{k=0}^{\infty} L^{(-2^k)}(\lambda).$$

On any compact subset of the disc,

$$|L^{(n)} - 1| \leq K \cdot \frac{1}{|n|}, \qquad n \neq 0.$$

Applying this with $n = -2^k$, $k = 0, 1, 2, \ldots$, we see that the infinite product above converges uniformly on compact subsets of the disc to an analytic function h. Since $|L^{(n)}| \leq 1$, we also have $|h(\lambda)| \leq 1$. This function h will show us that ψ is a homeomorphism. The claim is that

$$\hat{h}(\psi(\lambda)) = \lambda.$$

If we restrict our attention to a fixed compact subset of D and use the inequality

thereon, we have
$$|L^{(n)} - 1| \leq K \cdot \frac{1}{|n|}$$

$$|h(L^{(2^n)}(\lambda)) - \lambda| = |L^{(2^n)}(\lambda) \prod_{k=0}^{\infty} L^{(2^n - 2^k)}(\lambda) - \lambda|$$

$$= |\lambda| \, |L^{(2^n)}(\lambda) \prod_{k=0}^{n-1} L^{(2^n - 2^k)}(\lambda) \prod_{k=n+1}^{\infty} L^{(2^n - 2^k)}(\lambda) - 1|$$

$$\leq |\lambda| \left[|L^{(2^n)}(\lambda) - 1| + \sum_{k=0}^{n-1} |L^{(2^n - 2^k)}(\lambda) - 1| + \sum_{k=n+1}^{\infty} |L^{(2^n - 2^k)}(\lambda) - 1| \right]$$

$$\leq |\lambda| \cdot K \left[\frac{1}{2^n} + \sum_{k=0}^{n-1} (2^n - 2^k)^{-1} + \sum_{k=n+1}^{\infty} (2^k - 2^n)^{-1} \right]$$

$$\leq |\lambda| \cdot K \left[2^{-n} + \sum_{k=0}^{n-1} 2^{1-n} + \sum_{k=n+1}^{\infty} 2^{1-k} \right]$$

$$\leq |\lambda| \cdot K \left[\frac{n+2}{2^{n-1}} \right]$$

$\to 0$ as $n \to \infty$.

The step from line 3 to line 4 in the preceding equation used the inequality

$$\left|1 - \prod_{k=1}^{\infty} \lambda_k \right| \leq \sum_{k=1}^{\infty} |\lambda_k - 1|$$

for any sequence of points in D. From $\hat{h} \circ \psi = z$ it is clear that ψ is a homeomorphism. The inverse of ψ is the restriction of $\hat{h}$ to the range of ψ.

Statement (iii), that $f \to \hat{f} \circ \psi$ maps H^∞ onto H^∞, is now perfectly clear. For if f is any function in H^∞, set $g = f \circ h$, and then

$$\hat{g} \circ \psi = f \circ \hat{h} \circ \psi = f.$$

Let us see what we have. The map ψ is a homeomorphism of the disc D into the fiber $\mathfrak{M}_1$. The set $\psi(D)$ is then topologically a disc. We can use ψ to endow $\psi(D)$ with an analytic structure, by just transferring the analytic structure of D. If we do this, then for any f in H^∞ the restriction of the representing function $\hat{f}$ to $\psi(D)$ is a bounded analytic function on the disc $\psi(D)$. Furthermore, this restriction map is onto all bounded analytic functions on $\psi(D)$. Let

$$B = \hat{H}^\infty |_{\psi(D)}.$$

Roughly, one might say that $B = H^\infty(\psi(D))$. In any event, B is a uniformly closed algebra of continuous functions on $\psi(D)$ which is isomorphic to H^∞. If we use the sup norm, B is then a commutative Banach algebra with identity. The maximal ideal space of B is

$$\mathfrak{M}(B) = \{\phi \in \mathfrak{M}(H^\infty); \; \phi(f) = 0 \text{ whenever } \hat{f} = 0 \text{ on } \psi(D)\}.$$

For, since B is a homomorphic image of H^∞, we can identify the complex homomorphisms of B with those complex homomorphisms of H^∞ which are "well-defined" on B. The maximal ideal space of B is, therefore, a subset of the fiber $\mathfrak{M}_1$. Since B is isomorphic to H^∞, we conclude that $\mathfrak{M}_1$ contains a homeomorphic replica of the entire maximal ideal space $\mathfrak{M}(H^\infty)$. Each of the fibers attached to $\psi(D)$ will contain a disc, attached to which are fibers containing discs, and so on. Thus the maximal ideal space $\mathfrak{M}$ reproduces itself ad infinitum inside each of the fibers $\mathfrak{M}_\alpha$.

Perhaps we should point out the fact that the map $f \to \hat{f} \circ \psi$ which we have been discussing is a norm-decreasing algebra homomorphism of H^∞ onto H^∞ which carries every function with continuous boundary values into a constant function. Of course, any function which is continuous at $\alpha = 1$ is carried onto a constant function.

L^∞ as a Banach Algebra

As usual, L^∞ denotes the space of bounded measurable (or Baire) functions on the unit circle, with complex values. If we identify functions which are equal almost everywhere with respect to Lebesgue measure, then L^∞ is a Banach space under the (Lebesgue) essential sup norm

$$||f|| = \text{ess sup } |f(e^{i\theta})|.$$

If we use pointwise multiplication, then L^∞ is a commutative Banach algebra with identity. The maximal ideal space of L^∞ is a totally (even extremally) disconnected compact Hausdorff space X, and L^∞ is isometrically isomorphic (via the Gelfand representation) to $C(X)$, the algebra of all continuous complex-valued functions on X. These facts have a variety of proofs. We shall present direct proofs, since they are easy and since their inclusion will provide us with suitable notation and terminology for what we do later.

Let $X = \mathfrak{M}(L^\infty)$, the space of complex homomorphisms of L^∞. We then have the Gelfand representation $f \to \hat{f}$ defined by $\hat{f}(\phi) = \phi(f)$.

Lemma. *For each* f *in* L^∞, $\sup_X |\hat{f}(\phi)| = ||f||$.

Proof. We know that $||\hat{f}||_\infty \leq ||f||$. To prove the reverse inequality, choose a complex number λ such that $|\lambda| = ||f||$ and the set $\{|f - \lambda| < \epsilon\}$ has positive measure for each $\epsilon > 0$. Such a λ can be found by the definition of the essential sup norm. Then $(f - \lambda)$ is not invertible in L^∞, because such an inverse would have modulus greater than $1/\epsilon$ on a set of positive measure, for each $\epsilon > 0$. Thus $(f - \lambda)L^\infty$ is a proper ideal in L^∞; it is contained in a maximal ideal, so there is a complex homomorphism ϕ of L^∞ such that

$$|\phi(f)| = |\lambda| = ||f||.$$

Lemma. *As* E *ranges over the measurable subsets of the unit circle, the open-closed sets*

$$\{\phi \in X; \ \hat{\chi}_E(\phi) = 0\}$$

give a basis for the topology of X. *In particular,* X *is totally disconnected.*

Proof. By definition, the sets of the form
$$U = \{|\hat{f}_j| < \epsilon, \ j = 1, \ldots, n\}$$
with $f_1, \ldots, f_n$ in L^∞ and $\epsilon > 0$ form a basis for the (weak) topology of X. The simple functions
$$\sum_{j=1}^{k} \lambda_j \chi_{E_j}$$
are dense in L^∞. From this it is easy to see that, in defining the basic open sets U above, we can restrict the f_j to being characteristic functions of measurable sets. But if E is a measurable set $\chi_E^2 = \chi_E$ so that $\hat{\chi}_E$ assumes only the values 0 and 1. Therefore, when $\epsilon < 1$
$$\{\hat{\chi}_{E_j} < \epsilon, j = 1, \ldots, n\} = \{\hat{\chi}_{E_j} = 0, j = 1, \ldots, n\}$$
$$= \{\hat{\chi}_E = 0\}$$
where E is the union of $E_1, \ldots, E_n$. That proves the lemma.

Theorem. *The Gelfand representation of* L^∞ *maps* L^∞ *isometrically and isomorphically onto* $C(X)$. *Also, the representation preserves complex conjugation.*

Proof. We know that the homomorphism $f \to \hat{f}$ is isometric, and hence is also an isomorphism. Let F be a continuous function on X. Since the open-closed sets $\{\hat{\chi}_E = 0\}$ form a basis for the topology of X, we can approximate F uniformly by functions of the form
$$\sum_{j=1}^{n} \lambda_j \hat{\chi}_{E_j}.$$
When we do so, the isometry of the Gelfand representation tells us that the simple functions $\sum \lambda_j \chi_{E_j}$ converge in L^∞ to a function f for which $\hat{f} = F$. Thus $\hat{L}^\infty = C(X)$. Approximation by simple functions makes it apparent that the conjugate of $\hat{f}$ is $(\bar{f})\hat{\ }$.

Corollary. *A subset* K *of* X *is both open and closed if and only if* K $= \{\hat{\chi}_E = 0\}$ *for some measurable set* E *on the unit circle.*

Proof. The set K is both open and closed if and only if the characteristic function of K is continuous. Since $f \to \hat{f}$ maps L^∞ onto $C(X)$, we see that K is open-closed if and only if $\chi_K = \hat{f}$ for some f in L^∞. Since $\chi_K^2 = \chi_K$, we have $f^2 = f$, and thus $f = \chi_E$ for some measurable set E.

We have seen that $X = \mathfrak{M}(L^\infty)$ is totally disconnected, i.e., it has a basis of open-closed sets. As we mentioned earlier, X is extremally disconnected, that is, every open subset of X has a closure which is also open. We shall have no need of this fact. It is quite easy to prove, by using the fact that every collection of measurable sets on the circle has a least upper bound. Given an open set U in K, let E denote the least upper bound of the measurable sets S for which $\{\hat{\chi}_S = 1\}$ is contained in U. Then $\{\hat{\chi}_E = 1\}$ is the closure of U.

The maximal ideal space of L^∞ has a natural projection onto the unit circle, much like the projection of the maximal ideal space of H^∞. If we call this projection σ, it is defined by

$$\sigma(\phi) = \phi(z), \qquad \phi \in X = \mathfrak{M}(L^\infty)$$

that is, $\sigma = \hat{z}$. Since $|z|^2 = 1$ on the unit circle, we have $|\hat{z}|^2 = 1$ on X. Therefore, σ maps X into the unit circle, and it is clear that the map is onto the circle. If α is a point on the unit circle, the set X_α, consisting of all complex homomorphisms of L^∞ which send z into α, will be called the **fiber of X over α:**

$$X_\alpha = \{\phi \in X = \mathfrak{M}(L^\infty); \; \phi(z) = \alpha\}$$
$$= \sigma^{-1}(\alpha).$$

Theorem. *Let f be a function in* L^∞. *The range of $\hat{f}$ on the fiber X_α consists of all complex numbers ζ with this property: for each neighborhood N of α and each $\epsilon > 0$, the set*

$$\{|f - \zeta| < \epsilon\} \cap N$$

has positive Lebesgue measure.

Proof. The number ζ fails to be in the range of $\hat{f}$ on X_α if and only if there is no proper ideal of L^∞ which contains both $(z - \alpha)$ and $(f - \zeta)$, i.e., if and only if there are functions g and h in L^∞ such that

$$(z - \alpha)g + (f - \zeta)h = 1.$$

This means simply that there exists a bounded measurable function h such that

$$\frac{1 - (f - \zeta)h}{z - \alpha}$$

is essentially bounded. Such an h exists if and only if $f - \zeta$ is essentially bounded away from zero in a neighborhood of α.

Corollary. *Let f be a function in* L^∞ *and let α be a point of the unit circle. Any value which $\hat{f}$ assumes on the fiber X_α is assumed on a non-empty open subset of X_α.*

Proof. Suppose $\hat{f}$ assumes the value 0 on X_α. Choose a neighborhood N_1 of α. The set $N_1 \cap \{|f| < 1\}$ has positive measure. We can, therefore, find a closed set E_1 on the unit circle such that E_1 has positive measure, E_1 does *not* contain α, and $|f| < 1$ on E_1. Let N_2 be a neighborhood of α which is disjoint from E_1. We can then find a closed set E_2 which is contained in N_2, has positive measure, does not contain α, and is such that

$|f| < \frac{1}{2}$ on E_2. We then choose a neighborhood N_3 of α and get a closed set E_3 of positive measure on which $|f| < \frac{1}{3}$, and which excludes α and is disjoint from E_1 and E_2. We continue in this way, making sure that the neighborhoods N_k satisfy

$$\bigcap_{k=1}^{\infty} N_k = \{\alpha\}.$$

We then have a sequence of pairwise disjoint closed subsets E_k of the unit circle such that

(i) the E_k converge to the point α.
(ii) $|f| < 1/k$ on E_k.

Now let E be the union of the E_k. We claim that $\hat{f}$ vanishes on the set $X_\alpha \cap \{\hat{\chi}_E = 1\}$ and that the latter set is non-empty. This set is non-empty by the last theorem, because the set where $\chi_E = 1$ has positive measure in every neighborhood of α. The function $\chi_E f$ tends continuously to 0 at α. By the theorem, $\hat{\chi}_E \hat{f}$ is identically 0 on X_α, i.e., $\hat{f}$ vanishes at every point of X_α where $\hat{\chi}_E = 1$. We are done.

This corollary is just a topological comment about the fibers X_α. It states that X_α is a compact Hausdorff space with the property that any continuous function F which vanishes at some point also vanishes on a non-empty open subset of X_α. Note that it does not assert that the set of zeros of F is open, but only that this set is either empty or has a non-empty interior.

The Šilov Boundary

We return now to the algebra H^∞, which we have represented by the algebra $\hat{H}^\infty$ of continuous functions on the compact maximal ideal space $\mathfrak{M} = \mathfrak{M}(H^\infty)$. The chief object of this section is to identify the Šilov boundary for H^∞ with the maximal ideal space of L^∞. First, let us make a few comments about "boundaries" for a function algebra.

Suppose that X is a compact Hausdorff space and that A is an algebra of continuous complex-valued functions on X which contains the constant functions and separates the points of X. A **boundary** for the algebra A is a subset S of X such that

$$\sup_X |f| = \max_S |f|, \quad \text{all } f \text{ in } A$$

that is, such that every function in A *attains* its maximum modulus on the set S. If S is closed, this just means that

$$\sup_X |f| = \sup_S |f|$$

for each f in A. Among the closed boundaries for A there is a smallest one. In other words, the intersection of all closed boundaries for A is a boundary for A. This smallest closed boundary is called the **Šilov boundary** for A. For a proof of its existence we refer the reader to [54; page 80]. If B is a commutative Banach algebra with identity, if X is the maximal ideal space of B, and if A is the representing algebra $\hat{B}$, we call this boundary the Šilov boundary for B.

The standard illustration of the concept of Šilov boundary is obtained when A is the algebra of continuous functions on the closed unit disc which are analytic in the interior. The maximum modulus theorem for analytic functions tells us that the Šilov boundary for A is a subset of the unit circle. It is very easy to see that it must be the entire unit circle, because A is rotation invariant; also, if $|\alpha| = 1$, then $f(z) = \frac{1}{2}(1 + \bar{\alpha}z)$ is a function in A which takes its maximum on the closed disc precisely at the point $z = \alpha$.

What can we say about the Šilov boundary for H^∞? Since we have embedded the unit disc as an open subset Δ of the space $\mathfrak{M} = \mathfrak{M}(H^\infty)$, the maximum modulus principle for analytic functions tells us that the Šilov boundary for H^∞ (i.e., for $\hat{H}^\infty$) is contained in the topological boundary of Δ in the space $\mathfrak{M}$. One's first guess might be that the Šilov boundary is all of $\bar{\Delta} - \Delta$; however, this is not the case. Here is an elementary proof. Let

$$f(z) = \exp\left(\frac{z+1}{z-1}\right)$$

so that f is in H^∞, f has no zeros in the open unit disc, and $f(z)$ tends to 0 as z approaches 1 along the positive axis. Therefore, the representing function $\hat{f}$ must have a zero somewhere on $\bar{\Delta} - \Delta$. No such zero of $\hat{f}$ can lie on the Šilov boundary. In fact, we must have $|\hat{f}| = 1$ on the Šilov boundary. Let U be the open subset of $\mathfrak{M}$ on which $|\hat{f}| < 1$. If U intersects the Šilov boundary, the minimality of this closed boundary tells us that there exists a function of g in H^∞ such that $|\hat{g}|$ does not attain to its maximum of the complement of U. By the definition of U, we then have $|\hat{f}\hat{g}| < \|g\|_\infty$ on all of $\mathfrak{M}$. But the boundary values of f are of absolute value 1 almost everywhere on the unit circle, so multiplication by f is an isometry. Thus, we have the contradiction

$$\|g\|_\infty > \|\hat{f}\hat{g}\|_\infty$$
$$= \|fg\|$$
$$= \|g\|.$$

We conclude that the Šilov boundary is a proper closed subset of $\bar{\Delta} - \Delta$. Later, we shall see that $\bar{\Delta} - \Delta$ is a great deal larger than the Šilov boundary, because the first set is connected and the second is totally disconnected.

To identify the Šilov boundary for H^∞, we proceed as follows. If we identify each function f in H^∞ with the function F in L^∞ determined by the boundary values of f, we may regard H^∞ as a closed subalgebra of L^∞ (the algebra of functions in L^∞ whose negative Fourier coefficients vanish). This provides us with a natural mapping τ from $X = \mathfrak{M}(L^\infty)$ into $\mathfrak{M} = \mathfrak{M}(H^\infty)$. This map τ is defined by restricting each complex homomorphism of L^∞ to the subalgebra H^∞. It is easy to see that τ is continuous.

Theorem. *The map τ is a homeomorphism of X into $\mathfrak{M}$, and the image set $\tau(X)$ is the Šilov boundary for H^∞; that is, $\tau(X)$ is the smallest closed subset of $\mathfrak{M}$ on which every $\hat{f}$ (f in H^∞) attains its maximum modulus.*

Proof. Since X is compact and τ is continuous, we can prove that τ is a homeomorphism by showing that τ is one-one. The latter means that the functions in H^∞ separate the complex homomorphisms of L^∞, and it will soon become apparent that this separation condition is satisfied.

Certainly, each $\hat{f}$, f in H^∞, attains its maximum modulus on the closed set $\tau(X)$, because, if F denotes the "boundary values" of f,

$$\sup_{\mathfrak{M}} |\hat{f}| = \|f\| = \|F\| = \sup_X |\hat{F}| = \sup_{\tau(X)} |\hat{f}|.$$

Thus $\tau(X)$ is a closed boundary for H^∞. Also, there is no proper closed subset of $\tau(X)$ on which all the $\hat{f}$ attain their maximum moduli. This fact can be proved as follows. Any proper closed subset of X omits an open-closed set U. Such a set U has the form

$$U = \{\phi \in X; \; \hat{\chi}_E(\phi) = 1\}$$

where E is some measurable subset of the circle. If u is the harmonic extension of χ_E to the disc, if v is a harmonic conjugate of u, and if $f = e^{u+iv}$, then f is in H^∞ and $|f| = \exp(\chi_E)$ almost everywhere on the circle. If F is the boundary function for f, we have $|\hat{F}| = \exp(\hat{\chi}_E)$ on X, or

$$|\hat{F}| = \begin{cases} e, & \text{on } U \\ 1, & \text{on } X - U. \end{cases}$$

Thus $|\hat{F}|$ does not attain its maximum on $X - U$. We conclude that there is no proper closed subset of X on which all the functions $\hat{F}$, representing H^∞ functions, attain their maximum moduli. The construction above shows also that the elements of H^∞ separate the complex homomorphisms of L^∞. Therefore X is homeomorphic to $\tau(X)$, and there is no proper closed subset of $\tau(X)$ on which all the $\hat{f}$, f in H^∞, attain their maximum moduli.

In the last few sections, we have discussed various restriction maps of one maximal ideal space into another. Let us summarize. Again, we let A denote the algebra of continuous functions on the circle whose negative Fourier coefficients vanish. We then have

$$A \subset H^\infty \subset L^\infty = C(X)$$

and

$$\sigma \begin{pmatrix} X \\ \downarrow \tau \\ \mathfrak{M}(H^\infty) \\ \downarrow \pi \\ \mathfrak{M}(A) \end{pmatrix}$$

Of course, $\mathfrak{M}(A)$ is the closed unit disc in the plane. We did not originally define π as the indicated restriction map, but rather as "evaluation at the function z." The maximal ideal space of A is identified as the closed disc by making the homomorphism ϕ in $\mathfrak{M}(A)$ correspond to the point $\phi(z)$. This identifies π with the restriction map from $\mathfrak{M}(H^\infty)$ into $\mathfrak{M}(A)$. In our diagram, we know the following:

(i) π is a homeomorphism over the open unit disc;
(ii) π is many-to-one over the points of the unit circle;
(iii) τ is a homeomorphism of X into $\mathfrak{M}(H^\infty)$;
(iv) $\tau(X)$ is a proper closed subset of $\bar{\Delta} - \Delta$, and is the Šilov boundary for H^∞;
(v) $\sigma = \pi \circ \tau$.

From this point on, *we shall identify X and $\tau(X)$*, and so regard X as a subset of $\mathfrak{M}(H^\infty)$.

Inner Functions and the Šilov Boundary

Suppose that f is an inner function in H^∞. Then $|f| = 1$ almost everywhere on the unit circle; thus it is apparent that the representing function $\hat{f}$ is of absolute value 1 on the Šilov boundary $X = \mathfrak{M}(L^\infty)$. In this section we shall prove D. J. Newman's theorem which shows, among other things, that this modular property characterizes the Šilov boundary. The result states that if ϕ is a complex homomorphism of H^∞, the following are equivalent: (i) ϕ lies on the Šilov boundary, i.e., ϕ extends to a complex homomorphism of L^∞; (ii) $|\phi(f)| = 1$ for every inner function f; (iii) $|\phi(B)| = 1$ for every Blaschke product B; (iv) $\phi(f) \neq 0$ for every inner function f; (v) $\phi(B) \neq 0$ for every Blaschke product B.

One can, of course, also phrase the result in terms of maximal ideals, rather than complex homomorphisms. Since a maximal ideal M in H^∞ lies on the Šilov boundary if and only if M is the intersection of H^∞ with a maximal ideal in L^∞, the important part of the result is that any maximal ideal in H^∞ which does not lie on the Šilov boundary contains a Blaschke product. As a first step, we prove a theorem which is of independent interest.

Theorem. *Every inner function is a uniform limit of Blaschke products.*

Proof. When we defined a Blaschke product B in Chapter 5, the function B was completely determined by its sequence of zeros. In the present context, we allow ourselves to multiply B by a constant of modulus 1 and still call the resulting function a Blaschke product.

Let f be an inner function. We wish to approximate f by Blaschke products. We may assume that $f(0) \neq 0$. If $|\alpha| < 1$, define

$$g_\alpha(z) = \frac{f(z) - \alpha}{1 - \bar{\alpha} f(z)}.$$

It is clear that g_α is an inner function. We shall find a sequence of numbers α_n such that $\alpha_n \to 0$ and each g_{α_n} is a Blaschke product. This will prove the theorem.

Define

$$J(r, \alpha) = \frac{1}{2\pi} \int_{-\pi}^{\pi} \log |g_\alpha(re^{i\theta})| d\theta.$$

g_α is a Blaschke product if and only if $\lim_{r \to 1} J(r, \alpha) = 0$; for, this limit is $-\int d\mu$, where μ is the positive singular measure on the circle which determines the singular part of g_α. Let

$$J(\alpha) = \lim_{r \to 1} J(r, \alpha).$$

Since we are only interested in small values of α, let us assume that $|\alpha| < |f(0)|$. Then $g_\alpha(0) \neq 0$, and so the convergence of $J(r, \alpha)$ to $J(\alpha)$ is bounded:

$$\log |g_\alpha(0)| \leq J(r, \alpha) \leq 0.$$

Suppose we consider a positive number $\rho < |f(0)|$. Then $J(r, \alpha)$ is uniformly bounded on $|\alpha| = \rho$. Consequently

$$\frac{1}{2\pi} \int_{-\pi}^{\pi} J(\rho e^{i\theta}) d\theta = \lim_{r \to 1} \frac{1}{2\pi} \int_{-\pi}^{\pi} J(r, \rho e^{i\theta}) d\theta$$

$$= \lim_{r \to 1} \frac{1}{2\pi} \int_{-\pi}^{\pi} \frac{1}{2\pi r} \int_{|z|=r} \log \left| \frac{f(z) - \rho e^{i\theta}}{1 - \rho e^{-i\theta} f(z)} \right| |dz| d\theta$$

$$= \lim_{r \to 1} \frac{1}{2\pi r} \int_{|z|=r} I_\rho(z) |dz|$$

where

$$I_\rho(z) = \frac{1}{2\pi} \int_{-\pi}^{\pi} \log \left| \frac{f(z) - \rho e^{i\theta}}{1 - \rho e^{-i\theta} f(z)} \right| d\theta.$$

Now

$$I_\rho(z) = \frac{1}{2\pi} \int_{-\pi}^{\pi} \log |f(z) - \rho e^{i\theta}| d\theta - \frac{1}{2\pi} \int_{-\pi}^{\pi} \log |e^{i\theta} - \rho f(z)| d\theta.$$

By Jensen's formula, the second integral is 0 and the first integral is

$$I_\rho(z) = \max (\log \rho, \log |f(z)|).$$

Since f is an inner function
$$\lim_{r \to 1} \log |f(re^{it})| = 0$$
for almost every t. Thus $I_\rho(re^{it})$ tends to 0 as $r \to 1$, for almost every value of t. But $I_\rho(re^{it})$ is bounded. We conclude that
$$\frac{1}{2\pi} \int_{-\pi}^{\pi} J(\rho e^{i\theta}) d\theta = 0$$
and since $J \leq 0$ it must be that $J = 0$ a.e. Therefore there is a value of α such that $|\alpha| = \rho$ and $J(\alpha) = 0$, i.e., g_α is a Blaschke product. Since we can find such an α for each $\rho < |f(0)|$, the proof is complete.

Theorem. *Let ϕ be a complex homomorphism of H^∞, and suppose that $\phi(B) \neq 0$ for every Blaschke product B. Then*

(i) $|\phi(B)| = 1$ *for every Blaschke product* B;
(ii) $|\phi(f)| = 1$ *for every inner function* f;
(iii) *for every* f *in* H^∞, *the number* $\phi(f)$ *is in the range of* $\hat{f}$ *on the Šilov boundary* $X = \mathfrak{M}(L^\infty)$.

Proof. (i) If ϕ does not annul any Blaschke product, ϕ is not evaluation at a point of the open unit disc. Thus $|\phi(z)| = 1$, and in fact $|\phi(B)| = 1$ for any finite Blaschke product B. Thus we only need prove that $|\phi(B)| = 1$ when B has the form
$$B(z) = \prod_{n=1}^{\infty} \left(\frac{\bar{\alpha}_n}{|\alpha_n|}\right)\left[\frac{\alpha_n - z}{1 - \bar{\alpha}_n z}\right], \quad 0 < |\alpha_n| < 1, \quad \Sigma (1 - |\alpha_n|) < \infty.$$
Choose a non-decreasing sequence of positive integers p_n such that
$$\lim_n p_n = \infty$$
$$\Sigma p_n(1 - |\alpha_n|) < \infty.$$
If N is a positive integer, define
$$\tilde{B}(z) = \prod_{n=1}^{\infty} [F_n(z)]^{p_n}$$
$$P_N(z) = \prod_{n=1}^{N} [F_n(z)]^{p_N - p_n}$$
$$Q_N(z) = \prod_{n=N+1}^{\infty} [F_n(z)]^{p_n - p_N}$$
where
$$F_n(z) = \frac{\bar{\alpha}_n}{|\alpha_n|}\left[\frac{\alpha_n - z}{1 - \bar{\alpha}_n z}\right].$$
Then
$$B^{p_N} Q_N = \tilde{B} P_N.$$
As we just noted, $|\phi(P_N)| = 1$, and certainly $|\phi(Q_N)| \leq 1$. Hence

$$|\phi(B)|^{p_N} \geq |\phi(\tilde{B})| > 0.$$

As $N \to \infty$ we have $p_N \to \infty$ and so $|\phi(B)| = 1$.

(ii) Since every inner function is a uniform limit of Blaschke products, and since ϕ is continuous, (ii) follows from (i).

(iii) Suppose f is a function in H^∞ such that $\hat{f}$ does not vanish anywhere on the Šilov boundary. Then $|f|$ is essentially bounded away from 0 on the unit circle. Thus, if $f = gF$ where g is inner and F is outer, $|F|$ is bounded away from 0 on the circle, and this implies that $1/F$ is a bounded analytic function. Hence $\phi(F) \neq 0$. We have

$$|\phi(f)| = |\phi(g)| \, |\phi(F)|$$
$$= |\phi(F)|$$
$$> 0.$$

That proves (iii).

Lemma. *Let ϕ be any complex homomorphism of H^∞. If u is a real-valued function in L^∞, define*

$$\ell(u) = \log |\phi(F_u)|$$

where F_u is the bounded analytic function

$$F_u(z) = \exp\left[\frac{1}{2\pi} \int_{-\pi}^{\pi} \frac{e^{i\theta} + z}{e^{i\theta} - z} u(e^{i\theta}) d\theta\right].$$

Then ℓ is a bounded and positive linear functional on the space of real-valued L^∞ functions. Furthermore, the function ψ defined by

$$\psi(f) = \ell(\mathrm{Re}\, f) + i\ell(\mathrm{Im}\, f)$$

is a linear functional of norm 1 on L^∞ whose restriction to H^∞ is ϕ.

Proof. First we show that ℓ is additive.

$$\ell(u_1 + u_2) = \log |\phi(F_{u_1+u_2})|$$
$$= \log |\phi(F_{u_1} F_{u_2})|$$
$$= \log |\phi(F_{u_1}) \phi(F_{u_2})|$$
$$= \ell(u_1) + \ell(u_2).$$

and so ℓ is additive.

It is clear that ℓ is bounded by 1; if $|u| \leq 1$ then

$$\frac{1}{e} \leq |F_u| \leq e$$

and (since F_u is invertible in H^∞) we have

$$\frac{1}{e} \leq |\phi(F_u)| \leq e$$

or

$$-1 \leq \ell(u) \leq 1.$$

Since ℓ is additive and bounded, it is continuous and hence linear.

The positivity of ℓ follows from the fact that $||\ell|| = \ell(1) = 1$. It can also be checked directly by the same sort of argument as above. If $u \geq 0$ then we have

$$1 \leq |F_u|$$
$$1 \leq |\phi(F_u)|$$
$$0 \leq \ell(u).$$

Since ℓ is a positive linear functional of norm 1, it is immediate that the complex extension ψ of ℓ, defined by

$$\psi(u + iv) = \ell(u) + i\ell(v)$$

is a positive linear functional of norm 1 on L^∞. If $f = u + iv$ is in H^∞, then $\psi(f) = \phi(f)$. It suffices to prove this when f vanishes at the origin. We then have

$$F_u = e^f$$
$$F_v = e^{-if}$$

and so

$$\begin{aligned}\psi(f) &= \ell(u) + i\ell(v) \\ &= \log|\phi(e^f)| + i\log|\phi(e^{-if})| \\ &= \log|e^{\phi(f)}| + i\log|e^{-i\phi(f)}| \\ &= \operatorname{Re} \phi(f) + i \operatorname{Im} \phi(f) \\ &= \phi(f).\end{aligned}$$

What the above Lemma does is exhibit an explicit Hahn-Banach (norm-preserving) extension of ϕ to a linear functional ψ on L^∞.

Theorem. *If ϕ is a complex homomorphism of H^∞, the following are equivalent.*

(i) *ϕ lies in the Šilov boundary for H^∞.*
(ii) *ϕ is the restriction to H^∞ of a complex homomorphism of L^∞.*
(iii) *$|\phi(f)| = 1$ for every inner function f.*
(iv) *$|\phi(B)| = 1$ for every Blaschke product B.*
(v) *$\phi(f) \neq 0$ for every inner function f.*
(vi) *$\phi(B) \neq 0$ for every Blaschke product B.*
(vii) *For every f in H^∞, the number $\phi(f)$ is in the range of $\hat{f}$ on the Šilov boundary for H^∞.*

Proof. We have previously done most of the work. In fact, we can complete the proof by showing that (vii) implies (ii). Suppose that ϕ has property (vii). We refer to the Lemma above. First we show that for any real-valued function u in L^∞ the number $\ell(u) = \log|\phi(F_u)|$ belongs to the range of $\hat{u}$ on the space $X = \mathfrak{M}(L^\infty)$. This is clear since $|\phi(F_u)|$ belongs to the range of $\hat{F}_u$ on X and $\log|\hat{F}_u| = \hat{u}$. We combine this observation with the fact that X is totally disconnected to show that the linear functional ℓ is simply evaluation at a point of X, i.e., there exists $x \in X$ such that $\ell(u) = \hat{u}(x)$ for all real values of u in L^∞. Since X is totally disconnected, finite linear combinations of idempotent functions (simple functions) are dense in L^∞. If no such point exists, then for each value of x in X there is an idempotent e in L^∞ such that $\hat{e}(x) = 1$ but $\ell(e) = 0$, because for each idempotent e the value of $\ell(e)$ must be 0 or 1. Since X is compact, we can find a finite number of idempotents $e_1, \ldots, e_n$ such that

$$\hat{e}_1 + \cdots + \hat{e}_n \geq 1$$
$$\ell(e_k) = 0, \quad k = 1, \ldots, n.$$

If we set $u = e_1 + \cdots + e_n$, then $\ell(u) = 0$ is not in the range of $\hat{u}$ on X.

Now we know that ℓ is evaluation at a point of X, that is, that ℓ is multiplicative. Therefore the complex extension ψ of ℓ is a complex homomorphism of L^∞ whose restriction to H^∞ is ϕ.

In connection with this Theorem, we should mention two facts which we shall see later. First, for any individual inner function f, the set of points in $\mathfrak{M}(H^\infty)$ where $|\hat{f}| = 1$ is always larger than the Šilov boundary. Second, a maximal ideal M off the Šilov boundary contains a Blaschke product B; however, it is not necessarily the case that M is in the closure of the sequence of zeros of B in the unit disc.

Representing Measures and Annihilating Measures

For a function algebra, one of the useful features of the Šilov boundary is that each complex homomorphism of the algebra can be represented as integration with respect to a positive measure on the boundary. In this section we shall discuss such "representing measures" for the algebra H^∞, and the results we obtain will give us some information about annihilating measures for H^∞.

Let Y be a compact Hausdorff space, and let A be a uniformly closed separating algebra of continuous functions on Y which contains the constant functions. Let X be the Šilov boundary for A, and let ϕ be a complex homomorphism of A. With the sup norm A is a commutative Banach algebra with identity, and so ϕ is necessarily continuous:

$$|\phi(f)| \leq \sup_Y |f|, \quad \text{all } f \text{ in } A.$$

The map $f \to f|_X$, which restricts each function f in A to the Šilov boundary, is a (sup) norm-preserving isomorphism of A with a subalgebra of $C(X)$. Thus, we may regard A as a subalgebra of $C(X)$, and ϕ is a bounded linear functional on this subalgebra:

$$|\phi(f)| \leq \sup_X |f|.$$

By the Hahn-Banach theorem, we can extend ϕ to a linear functional of bound 1 on $C(X)$. Such a functional is determined by a finite complex Baire measure m_ϕ on X, and so

$$\phi(f) = \int_X f\, dm_\phi$$

for every f in A. Now m_ϕ is a complex measure of total variation 1 and $\int dm_\phi = 1$. From this it is easy to see that m_ϕ must be a positive measure.

We conclude that for each complex homomorphism ϕ of A there exists at least one positive measure m_ϕ on the Šilov boundary X such that

$$\phi(f) = \int_X f\, dm_\phi, \quad \text{all } f \text{ in } A.$$

Any such positive measure we shall call a **representing measure** for ϕ. We notice that in the above argument we could replace X by any closed subset S of X such that

$$|\phi(f)| \leq \sup_S |f|, \quad f \text{ in } A.$$

Such a set we shall call a **support set for** ϕ, since it is a set which will support a representing measure for ϕ.

If we apply these results to H^∞, we conclude that for each complex homomorphism ϕ of H^∞ there is at least one positive measure m_ϕ on the Šilov boundary $X = \mathfrak{M}(L^\infty)$ such that

$$\phi(f) = \int_X \hat{f}\, dm_\phi, \quad f \text{ in } H^\infty.$$

As we shall soon see, each ϕ in $\mathfrak{M}(H^\infty)$ actually has a *unique* representing measure m_ϕ.

What sort of condition on an algebra might guarantee that representing measures are unique? If one has two positive measures m_ϕ and μ_ϕ on the Šilov boundary which represent the same homomorphism ϕ, then the difference $m_\phi - \mu_\phi$ is a real measure on the boundary which is orthogonal to the algebra. Thus, the most obvious condition which would guarantee uniqueness of representing measures for all homomorphisms is that no non-zero real measure on the Šilov boundary is orthogonal to the algebra. This just means that the algebra is a Dirichlet algebra on its Šilov boundary, i.e., that the real parts of the functions in the algebra are dense in the real continuous functions on the Šilov boundary. Now H^∞ is *not* a Dirichlet algebra on its Šilov boundary, that is, the real parts of the functions in H^∞ are not (uniformly) dense in the real L^∞ functions. We gave one proof of this in Chapter 9, where we showed that any function in the closure of $H^\infty + \bar{H}^\infty$ has a conjugate harmonic function of growth

$o\left(\log\dfrac{1}{1-r}\right)$ in the disc. There is a more abstract proof which is short, and worth presenting.

Theorem. *Let* X *be a totally disconnected compact Hausdorff space, and let* A *be a Dirichlet algebra on* X. *Then* A = C(X).

Proof. Let S be any open-closed subset of X and let χ_S be its characteristic function. Then χ_S is a real continuous function on X. Consequently there is a function f in A such that

$$|\chi_S - \operatorname{Re} f| < \frac{1}{2}$$

on X. The range of f does not intersect the line $\operatorname{Re} z = \frac{1}{2}$ in the plane. Hence we can find a sequence of polynomials $p_n(z)$ which converge uniformly to 1 on that part of the range of f to the right of the line $\operatorname{Re} z = \frac{1}{2}$ and which converge uniformly to 0 on that part of the range of f to the left of $\operatorname{Re} z = \frac{1}{2}$. Then $p_n(f)$ is a sequence of functions in A which converge uniformly to χ_S. Thus χ_S is in A. Since X is totally disconnected and A contains the characteristic function of every open-closed subset of X, it follows that $A = C(X)$.

If we apply this argument to H^∞, we conclude that H^∞ is not a Dirichlet subalgebra of L^∞; indeed, we see that the uniform closure of Re H^∞ does not contain the characteristic function of any measurable set E on the unit circle, unless E has (normalized) Lebesgue measure 0 or 1. In spite of this fact, representing measures are unique for every complex homomorphism of H^∞. This is because H^∞ possesses a property which is very close to the Dirichlet property: Every real-valued function u in L^∞ is the logarithm of the modulus of an invertible function F in H^∞:

$$F(z) = \exp\left[\frac{1}{2\pi}\int_{-\pi}^{\pi}\frac{e^{i\theta}+z}{e^{i\theta}-z}u(e^{i\theta})d\theta\right].$$

Theorem. *Let* X *be a compact Hausdorff space, and let* A *be a uniformly closed subalgebra of* C(X) *which contains the constant functions. Suppose that there is a dense subspace* S *of the real-valued continuous functions on* X *such that each* u *in* S *has the form* $u = \log|F_u|$, *where* F_u *is an invertible element of* A. *If* ϕ *is a complex homomorphism of* A, *then* ϕ *has a unique representing measure* m_ϕ *on* X. *Furthermore,* m_ϕ *is the measure defined by*

$$\int u\, dm_\phi = \log|\phi(F_u)|, \quad u \in S.$$

Proof. Let m and ρ be two (positive) measures on X, each of which represents ϕ. Let u be a function in S, $u = \log|F_u|$, where F_u is an invertible function in A. Since

$$\phi(F_u) = \int_X F_u\, dm$$

and
$$\phi(F_u^{-1}) = \int_X F_u^{-1} d\rho$$
we have
$$|\phi(F_u)| \leq \int e^u dm$$
$$|\phi(F_u^{-1})| \leq \int e^{-u} d\rho.$$
But $\phi(F_u)\phi(F_u^{-1}) = 1$, and so
$$\int e^u dm \geq [\int e^{-u} d\rho]^{-1}.$$
Since S is a subspace, the same inequality holds with u replaced by tu, where t is any real number. Therefore
$$\int \left(1 + tu + \frac{t^2}{2!} u^2 + \cdots\right) dm \geq \left[\int \left(1 - tu + \frac{t^2}{2!} u^2 - \cdots\right) d\rho\right]^{-1}$$
for every real value of t, that is
$$1 + t \int u\, dm + \frac{t^2}{2!} \int u^2 dm + \cdots \geq 1 + t \int u\, d\rho + \cdots$$
for all real values of t. Clearly then
$$\int u\, dm \geq \int u\, d\rho.$$
Since this is true for every u in S, which is dense in $C_R(X)$, we conclude that $\rho \leq m$. Since ρ and m are positive measures of mass 1, it follows that $\rho = m$.

If u is in S define $\ell(u) = \log |\phi(F_u)|$, where F_u is any invertible function in A such that $\log |F_u| = u$. It is clear that ℓ is a well-defined function on S, which is additive because ϕ is multiplicative. Also
$$\ell(u) = \log |\phi(F_u)|$$
$$\leq \log ||F_u||$$
$$= \log ||e^u||$$
$$\leq \max_X u.$$
Therefore $|\ell(u)| \leq ||u||$, i.e., ℓ is bounded by 1. Since ℓ is additive and bounded it is continuous and linear on S. There is a unique extension of ℓ to a bounded linear functional on $C_R(X)$, and the extended functional is given by a positive measure m_ϕ on X. To see that m_ϕ represents ϕ we need only show that $\int u\, dm_\phi = \operatorname{Re} \phi(f)$ when $u = \operatorname{Re} f$ with f in A. Since $u = \log |e^f|$
$$\int u\, dm_\phi = \ell(u)$$
$$= \log |\phi(e^f)|$$
$$= \log |e^{\phi(f)}|$$
$$= \operatorname{Re} \phi(f).$$
That completes the proof.

Another way to state the uniqueness part of this last theorem is: If a linear functional ϕ on A happens to be multiplicative, then there is a unique Hahn-Banach (norm-preserving) extension of ϕ to a linear functional on $C(X)$. One should note that the hypothesis of this theorem is satisfied if A is a Dirichlet algebra on the space X. For, the space $S = \operatorname{Re} A$ has the stated property: If $u = \operatorname{Re} f$ with f in A then $u = \log |e^f|$. As we said, the hypothesis is also satisfied when $X = \mathfrak{M}(L^\infty)$ and A is the algebra $\hat{H}^\infty$. Here we can in fact take S as the space of all real-valued continuous functions on X. Thus, each complex homomorphism ϕ of H^∞ has a unique representing measure m_ϕ on $X = \mathfrak{M}(L^\infty)$. It is defined as follows: If u is a real L^∞ function then

$$\int_X \hat{u}\, dm_\phi = \log |\phi(F_u)|$$

where F_u is any outer function in H^∞ such that $\log |F_u| = u$ almost everywhere on the unit circle. For the case in which ϕ is evaluation at a point λ in the open unit disc, the uniqueness of m_ϕ was proved by Gleason and Whitney. In this case the description of m_ϕ reduces to

$$\int_X u\, dm_\phi = \frac{1}{2\pi}\int_{-\pi}^{\pi} u(e^{i\theta}) P_\lambda(\theta)\, d\theta$$

where P_λ is the Poisson kernel for the point λ. In particular, when ϕ is evaluation at the origin, the unique Hahn-Banach extension of ϕ to a linear functional ψ on L^∞ is given by

$$\psi(f) = \frac{1}{2\pi}\int_{-\pi}^{\pi} f(e^{i\theta})\, d\theta, \quad f \in L^\infty.$$

If we combine our last result with the methods of Helson and Lowdenslager (discussed in Chapter 4) we can obtain results about the relation of representing measures for H^∞ to annihilating measures for H^∞ and also to arbitrary positive measures on the Šilov boundary. It is not surprising that we can say more about representing measures for points in the open unit disc, but the first result we should discuss is Szegö's theorem, which is valid in the context of the last Theorem.

Theorem. *Let* A *be a uniformly closed algebra of continuous complex-valued functions on the compact Hausdorff space* X *($1 \in$ A). Suppose that the class of functions* $\log |F|$, *where* F *is an invertible element of* A, *contains a dense subspace* S *of the real-valued continuous functions on* X. *Let* m *be any positive measure on* X *which is multiplicative on* A, *and let* A_0 *be the set of functions* f *in* A *such that* $\int f\, dm = 0$. *Then for every positive measure* μ *on* X

$$\inf_{f \in A_0} \int |1-f|^2 d\mu = \exp\left[\int \log\left(\frac{d\mu}{dm}\right) dm\right].$$

Proof. We are assuming that m is a representing measure for a complex homomorphism ϕ of A. As we proved in the last theorem, m is the unique representing measure for ϕ, and is defined by

$$\int_X u\, dm = \log |\phi(F_u)|$$

when $u \in S$ and F_u is an invertible element of A such that $u = \log |F_u|$. From the specific form of m we can easily show that m satisfies Jensen's inequality:

$$\log |\phi(f)| = \log |\textstyle\int f\, dm| \leq \textstyle\int \log |f|\, dm, \quad \text{for all } f \in A.$$

Given f in A, let $\epsilon > 0$. Then $\log (|f| + \epsilon)$ is a continuous function on X. Choose $u = \log |F_u|$ in the subspace S so that

$$u - \epsilon < \log (|f| + \epsilon) < u + \epsilon.$$

Then $|fF_u^{-1}| < e^\epsilon$ on X. Therefore

$$|\phi(f)\, \phi(F_u^{-1})| < e^\epsilon.$$

If we take the logarithm of both sides of this inequality we have

$$\log |\phi(f)| - \textstyle\int u\, dm < \epsilon.$$

But $u < \epsilon + \log (|f| + \epsilon)$ on X, and if we integrate with respect to m

$$\textstyle\int u\, dm < \epsilon + \textstyle\int \log (|f| + \epsilon)\, dm.$$

Hence

$$\log |\phi(f)| < 2\epsilon + \textstyle\int \log (|f| + \epsilon)\, dm$$

and if we let $\epsilon \to 0$, we have the Jensen inequality.

Now let μ be any positive measure on X, and let

$$I(\mu) = \inf_{f \in A_0} \textstyle\int |1 - f|^2\, d\mu.$$

If $d\mu = h\, dm + d\mu_s$ is the Lebesgue decomposition of μ relative to m, then $I(\mu) = I(\mu_a)$ where $d\mu_a = h\, dm$, the absolutely continuous part of μ. This follows from the uniqueness of m by the Helson-Lowdenslager argument which we presented in part (i) of the Theorem on page 44 of Chapter 4 and its Corollary 1. This reduces the problem to the case when $d\mu = h\, dm$ with h a non-negative function in $L^1(dm)$. In this case we wish to show that

$$I(\mu) = \exp \left(\textstyle\int \log h\, dm\right).$$

If f is in A_0, then by the Jensen inequality

$$\begin{aligned}
\textstyle\int |1 - f|^2 h\, dm &\geq \exp \left[\textstyle\int \log (|1 - f|^2 h)\, dm\right] \\
&= \exp \left[\textstyle\int \log |1 - f|^2\, dm\right] \exp \left[\textstyle\int \log h\, dm\right] \\
&\geq \exp \log |\phi(1 - f)|^2 \cdot \exp \left[\textstyle\int \log h\, dm\right] \\
&= 1 \cdot \exp \left[\textstyle\int \log h\, dm\right].
\end{aligned}$$

Therefore

$$I(\mu) \geq \exp\left[\int \log h\, dm\right].$$

Since m is a positive measure of mass 1

$$\exp\left[\int \log h\, dm\right] = \inf_g \int h e^g\, dm, \quad g = \bar{g} \in L^1(dm), \quad \int g\, dm = 0.$$

See the proof on page 48. By a simple argument, we need only employ continuous functions g in computing the last infimum (see page 49). Thus

$$\exp\left[\int \log h\, dm\right] = \inf \int h e^g\, dm, \quad g \in C_R(X), \quad \int g\, dm = 0.$$

If $g \in C_R(X)$ we can uniformly approximate g by functions $u = \log |F_u|^2$, where F_u is an invertible element of A. If $\int g\, dm = 0$, we can choose the approximating functions u so that $\phi(F_u) = 1$. This only involves replacing F_u by $[\phi(F_u)]^{-1} F_u$. Therefore

$$\inf_g \int h e^g\, dm = \inf_F \int h |F|^2\, dm, \quad \phi(F) = 1, \quad F, F^{-1} \in A.$$

If $\phi(F) = 1$ then $F = 1 - f$ where $f \in A_0$. Thus

$$\exp\left[\int \log h\, dm\right] = \inf_F \int h |F|^2\, dm$$

$$\geq \inf_{f \in A_0} \int h |1 - f|^2\, dm$$

$$= I(\mu).$$

This completes the proof.

Of course the above Szegö theorem holds when $X = \mathfrak{M}(L^\infty)$, $A = \hat{H}^\infty$, and m is the representing measure for any complex homomorphism of H^∞. For measures representing points in the open unit disc we also have an F. and M. Riesz theorem.

Theorem. *Let m_0 be the unique positive measure on $X = \mathfrak{M}(L^\infty)$ which represents the homomorphism "evaluation at the origin" on H^∞, and let $\hat{H}_0^\infty$ be the set of functions $\hat{f}$ where f is in H^∞ and vanishes at the origin. Let μ be a finite complex Baire measure on X which is orthogonal to $\hat{H}_0^\infty$. Then the absolutely continuous and singular parts of μ with respect to m_0 are separately orthogonal to $\hat{H}_0^\infty$, and the singular part is also orthogonal to 1.*

Proof. The proof is identical with the proof of the general F. and M. Riesz theorem in Chapter 4 (page 47). All that is required is: (i) m_0 is multiplicative on $\hat{H}^\infty$; (ii) any positive measure on X which agrees with m_0 on $\hat{H}^\infty$ is equal to m_0; (iii) the real parts of the functions in $\hat{H}^\infty$ are dense in the space of real functions in $L^2(dm_0)$. Hypothesis (iii) simply states that the real parts of the functions in H^∞ are dense in the real functions in L^2 of the circle.

Corollary. *If μ is a (complex) measure on X which is orthogonal to $\hat{H}^\infty$, and if μ is mutually singular with m_0, then μ is also orthogonal to all $\hat{f}$ where f is in the closed subalgebra of H^∞ which is generated by H^∞ and $\bar{z}$.*

Proof. This follows from the last theorem by an argument similar to one used in proving the F. and M. Riesz theorem in Chapter 4. Since μ is singular and orthogonal to $\hat{H}^\infty$, the singular measure $\bar{z}d\mu$ is orthogonal to $\hat{H}_0^\infty$, and hence is orthogonal to 1. Thus $(\bar{z}^2)\hat{\ }d\mu$ is orthogonal to $\hat{H}_0^\infty$, etc.

Corollary. *Let μ be any finite real measure on X which is orthogonal to $\hat{H}^\infty$. Then μ is mutually singular with m_0, and μ is, therefore, orthogonal to $\hat{f}$ for every function f which is in the closed subalgebra of L^∞ generated by H^∞ and $\bar{z}$.*

Proof. Let $d\mu = h dm_0 + d\mu_s$ be the Lebesgue decomposition of μ relative to m_0. Then the measures $h dm_0$ and $d\mu_s$ are separately orthogonal to $\hat{H}^\infty$; each of these measures is real, since μ is a real measure. Thus h is a real-valued function in $L^1(dm)$, and is orthogonal to the functions in $\hat{H}^\infty$ and their complex conjugates. If we use, for example, Cesaro means, it is then easy to see that h is orthogonal to $\hat{L}^\infty$. Thus $h = 0$ almost everywhere dm_0.

We now know that μ is a singular measure, and we can apply the previous corollary.

Algebras on the Fibers

Let α be a point on the unit circle. We denote by A_α the algebra obtained by restricting $\hat{H}^\infty$ to the fiber $\mathfrak{M}_\alpha$ in the maximal ideal space $\mathfrak{M} = \mathfrak{M}(H^\infty)$. In this section we shall study A_α, principally because it is a very interesting function algebra, but also because through the study of A_α we will obtain some results about H^∞.

Theorem. *(i) A_α is a uniformly closed subalgebra of $C(\mathfrak{M}_\alpha)$. (ii) The maximal ideal space of A_α is $\mathfrak{M}_\alpha$. (iii) The Šilov boundary for A_α is contained in $X_\alpha = X \cap \mathfrak{M}_\alpha$.*

Proof. Let $f(z) = \frac{1}{2}(1 + \bar{\alpha}z)$. As we have noted several times, f is a function in H^∞ such that $\hat{f} = 1$ on the fiber $\mathfrak{M}_\alpha$ and $|\hat{f}| < 1$ on the remainder of $\mathfrak{M}$. The three statements of this theorem all follow from the existence of this function f. (i) Since $\hat{H}^\infty$ is uniformly closed on $\mathfrak{M}$, and since $\mathfrak{M}_\alpha$ is the set where a function of norm 1 is equal to 1, the restriction of $\hat{H}^\infty$ to $\mathfrak{M}_\alpha$ is uniformly closed. The algebra A_α is isomorphic to $\hat{H}^\infty/I_\alpha$, where I_α is the ideal of functions in $\hat{H}^\infty$ which vanish on $\mathfrak{M}_\alpha$. Since I_α is closed, A_α inherits the complete quotient norm on the quotient of the Banach spaces $\hat{H}^\infty$ and I_α. If one uses the function $\hat{f}$, it is not difficult to show that the quotient norm on A_α is equal to the sup norm over $\mathfrak{M}_\alpha$. For the details, we refer the reader to the proof of Rudin's extension theorem in Chapter 6. (ii) Each complex homomorphism of A_α induces a complex homomorphism of H^∞, by composing with the restriction map. Thus one may identify the complex homomorphisms of A_α with those complex homomorphisms ϕ of H^∞ which are well-defined on A_α, i.e., those ϕ such

that $\phi(g) = 0$ whenever $\hat{g} = 0$ on $\mathfrak{M}_\alpha$. But these homomorphisms are exactly those in the fiber $\mathfrak{M}_\alpha$, because $1 - \hat{f}$ vanishes precisely on $\mathfrak{M}_\alpha$.
(iii) Let ϕ be a complex homomorphism of A_α; i.e., let ϕ be in $\mathfrak{M}_\alpha$. Let m_ϕ be the representing measure for ϕ on the Šilov boundary X for H^∞. If we use our distinguished function f, we see that

$$1 = \int_X (\hat{f})^n dm_\phi, \qquad n = 1, 2, 3, \ldots.$$

But as n increases, $(\hat{f})^n$ converges boundedly and pointwise to the characteristic function of the set $\mathfrak{M}_\alpha$. Thus

$$\int_{X \cap M_\alpha} dm_\phi = 1$$

and m_ϕ must be supported on X_α. In particular,

$$|\hat{g}(\phi)| \leq \sup_{X_\alpha} |\hat{g}|, \quad g \text{ in } H^\infty$$

so the Šilov boundary for A_α is a subset of X_α.

Corollary. *If ϕ is a complex homomorphism of H^∞ which is contained in the fiber $\mathfrak{M}_\alpha$, then the representing measure for ϕ is supported on X_α.*

Corollary. *Each fiber $\mathfrak{M}_\alpha$ is connected.*

Proof. A theorem of Šilov [82], which we shall not prove, states that if the maximal ideal space of a function algebra is disconnected, then the algebra contains a non-trivial idempotent function. Since we know that $\mathfrak{M}_\alpha$ is the maximal ideal space of A_α, we can prove that $\mathfrak{M}_\alpha$ is connected by showing that A_α contains no non-trivial idempotents. Thus we shall show that if g is in H^∞ and the function $(\hat{g})^2 - \hat{g}$ vanishes on $\mathfrak{M}_\alpha$, then either $\hat{g}$ is identically 0 on $\mathfrak{M}_\alpha$ or $\hat{g}$ is identically 1 on $\mathfrak{M}_\alpha$. We proved earlier that the range of $\hat{g}$ on the fiber $\mathfrak{M}_\alpha$ consists of all complex numbers λ for which there exists a sequence of points z_n in the open unit disc with $\lim z_n = \alpha$ and $\lim g(z_n) = \lambda$. From this it is clear that if the range of $\hat{g}$ on $\mathfrak{M}_\alpha$ consisted precisely of the numbers 0 and 1, the function g would map every sufficiently small neighborhood of α onto a disconnected set. This is (of course) absurd.

Corollary. *In the maximal ideal space of H^∞, the complement of the open unit disc is connected.*

Proof. Suppose $\mathfrak{M} - \Delta$ is not connected. Then $\mathfrak{M} - \Delta$ is the union of two non-empty disjoint closed sets, K_0 and K_1. Each fiber $\mathfrak{M}_\alpha$ is then the union of the disjoint closed sets $K_0 \cap \mathfrak{M}_\alpha$ and $K_1 \cap \mathfrak{M}_\alpha$. Since $\mathfrak{M}_\alpha$ is connected, one of these two sets is empty. The unit circle is, therefore, disconnected by the closed sets

$$\{\alpha; \mathfrak{M}_\alpha \subset K_0\} \quad \text{and} \quad \{\alpha; \mathfrak{M}_\alpha \subset K_1\}.$$

Now we want to prove what is perhaps the most striking property of A_α. In order to do so, we shall need the following definitions.

Let A be a collection of continuous complex-valued functions on a topological space S. We say that A is **regular on** S if, for each closed set K in S and each point p not in K, there is a function g in A such that $g = 0$ on K but $g(p) \neq 0$. We say that A is **normal on** S if, for each pair of disjoint closed sets K_0 and K_1 in S, there exists a function g in A such that $g = 0$ on K_0 and $g = 1$ on K_1.

Theorem. *The algebra* A_α *is regular on the space* X_α; *in fact, the collection of functions* $\{|F|; F \text{ in } A_\alpha\}$ *is normal on* X_α.

Proof. Recall that X_α is that part of the Šilov boundary X which lies in the fiber $\mathfrak{M}_\alpha$. We know that the Šilov boundary for A_α is contained in X_α, so we can reasonably identify A_α with its restriction to X_α. Since X is totally disconnected, so is X_α. What we shall prove is that if K is any (relatively) open-closed subset of X_α, there exists a function h in H^∞ such that $\hat{h} = 0$ on K and $|\hat{h}| = 1$ on $X_\alpha - K$.

For convenience, assume that $\alpha = 1$. Let K be an open-closed subset of X_1. Since X is totally disconnected, K is the intersection with X_1 of an open-closed subset of X. In other words, there is a measurable set E on the unit circle such that

$$K = \{\phi \in X_1; \hat{\chi}_E(\phi) = 0\}.$$

Let

$$u(e^{i\theta}) = [1 - \chi_E(e^{i\theta})] \log|1 - e^{i\theta}|.$$

Then u is in L^1 of the circle and is bounded above. Hence

$$h(z) = \exp\left[\frac{1}{2\pi}\int_{-\pi}^{\pi} \frac{e^{i\theta} + z}{e^{i\theta} - z} u(e^{i\theta})d\theta\right]$$

is in H^∞. Almost everywhere on the unit circle, we have

$$|h| = e^u = \begin{cases} 1, & \text{on } E \\ |1 - e^{i\theta}|, & \text{off } E. \end{cases}$$

From this we see that the function f defined by

$$f(e^{i\theta}) = \frac{[1 - \chi_E(e^{i\theta})]h(e^{i\theta})}{1 - e^{i\theta}}$$

is a bounded measurable function. Thus

$$h = (1 - z)f + \chi_E h$$

where f is in L^∞. Let ϕ be any complex homomorphism of L^∞ which lies in the fiber X_1 and for which $\phi(\chi_E) = 0$. Then

$$\phi(h) = \phi(1 - z)\phi(f) + \phi(\chi_E)\phi(h) = 0 \cdot \phi(f) + 0 \cdot \phi(h) = 0.$$

Thus $\hat{h}$ vanishes on the set K. But

$$[1 - |h|]\chi_E = 0 \quad \text{a.e.}$$

so that $|\hat{h}| = 1$ at any point of X where $\hat{\chi}_E = 1$. In particular, we have $\hat{h} = 0$ on K, and $|\hat{h}| = 1$ on $X_1 - K$.

Most of the remaining properties of A_α which we wish to discuss are consequences of this result. Two of these properties of A_α are (i) X_α is the Šilov boundary for A_α; (ii) A_α is contained in no maximal closed subalgebra of $C(X_\alpha)$. We shall prove these results for any function algebra A which is regular on a compact Hausdorff space X. In view of this, one may ask what makes A_α anything special as a function algebra. The answer is that A_α is, to the best of my knowledge, the first known example of a *uniformly closed* algebra which is regular on the compact space X but is not all of $C(X)$. It was also the first known function algebra not contained in a maximal algebra.

Theorem. *Let* X *be a compact Hausdorff space and let* A *be a uniformly closed algebra of continuous complex-valued functions on* X, *such that* A *separates the points of* X *and contains the constant functions. Suppose that* A *is regular on* X. *Then*

(i) X *is the Šilov boundary for* A;

(ii) *any function in* A *which vanishes on a non-empty open subset of* X *also vanishes on a non-empty open subset of the maximal ideal space of* A;

(iii) *if* ϕ *is a complex homomorphism of* A *and* K *is a minimal support set for* ϕ, *and if* f *is a function in* A *which vanishes on an open subset of* X *which intersects* K, *then* $\phi(f) = 0$;

(iv) X *is the maximal ideal space of* A *if and only if* A *is normal on* X;

(v) *if* K *is a closed subset of* X *with the property that the restriction of* A *to* K *is not dense in* C(K), *then* K *has a proper closed subset* S *such that the restriction of* A *to* S *is not dense in* C(S);

(vi) A *is contained in no maximal proper closed subalgebra of* C(X).

Proof. (i) This statement is a well-known (and obvious) fact. Let K be the Šilov boundary for A. If K were not all of X, there would exist a function in A which vanished on K but was not identically 0. (ii) Let f be a function in A which vanishes on a non-empty open set U in X. Then there exists a non-empty open set V whose closure is contained in U. Since A is regular on X, there is a function g in A which vanishes on the complement of V but is not identically zero. Now $fg = 0$. Let

$$N = \{\phi \in \mathfrak{M}(A); \phi(g) \neq 0\}.$$

Then N is a non-empty open subset of $\mathfrak{M}(A)$ and $\phi(f) = 0$ for each ϕ in N. (iii) Recall that a support set for the complex homomorphism ϕ is a closed subset K of X such that

$$|\phi(f)| \leq \sup_K |f|, \quad \text{all } f \in A.$$

Suppose that K is a minimal support set for ϕ, i.e., that no proper closed subset of K is a support set for ϕ. Let f be a function in A which vanishes on an open subset U of X such that $U \cap K$ is non-empty. Then $\phi(f) = 0$; for, suppose $\phi(f) \neq 0$. Then let m_ϕ be a representing measure for ϕ which is supported on K. Then

$$\phi(g) = \frac{1}{\phi(f)} \int_K gf dm_\phi$$

for every g in A. Let $d\mu = \frac{1}{\phi(f)} f dm_\phi$. Then μ is a finite complex measure whose closed support is contained in $K - U$, and

$$\phi(g) = \int g d\mu$$

for all g in A. Then there is a constant k such that

$$|\phi(g)| \leq k \sup_{K-U} |g|$$

for every g in A. This means that ϕ is a bounded complex homomorphism of the restriction algebra $A|_{K-U}$. Hence ϕ extends to a complex homomorphism of the uniform completion of this restriction algebra. This completion is a commutative Banach algebra with identity, so any complex homomorphism of it is automatically of norm 1. Thus

$$|\phi(g)| \leq \sup_{K-U} |g|, \quad g \text{ in } A$$

and $K - U$ is a support set for ϕ. This contradicts the minimality of K. [We should remark that the proof of (iii) made no use of the regularity of A.] (iv) It is well-known that a commutative Banach algebra which is regular on its maximal ideal space is actually normal thereon. We refer the reader to [54; page 84] for a proof. We are chiefly interested here in the converse, which is due to Bishop. Suppose that A is normal on X. We shall prove that X is the maximal ideal space of A. Let ϕ be any complex homomorphism of A. We shall prove that ϕ is evaluation at a point of X. Let K be a minimal support set for ϕ. The existence of such a set is a simple consequence of Zorn's lemma. Suppose that K contains at least two points. Then we can find two closed subsets K_0 and K_1 of X such that the *interior* of each has a non-empty intersection with K. Since A is normal on X, there exists a function f in A such that $f = 0$ on K_0 and $f = 1$ on K_1. By (iii), we must have $\phi(f) = 0$ and $\phi(1 - f) = 0$, an absurdity. We conclude that K is a single point, and hence that ϕ is evaluation at that point. (v) Suppose that $A|_K$ is not dense in $C(K)$. Then there is a finite complex Baire measure μ on the set K such that $\mu \neq 0$ and μ annihilates A:

$$\int_K f d\mu = 0, \quad f \text{ in } A.$$

Let x be a point in the closed support of μ. The support of μ must contain more than one point; hence, there is an open subset U of X which intersects the support of μ and whose closure does not contain x. Let f be a function in A such that $f = 0$ on U but $f(x) \neq 0$. Let $d\mu_1 = f d\mu$, and μ_1 is a measure which annihilates A. Also, μ_1 is supported on the closed set $S = K - U$, a proper subset of K. Since x is in the support of μ and $f(x) \neq 0$, we cannot have $\mu_1 = 0$. We have a non-zero measure on S which annihilates A, so the restriction of A to S is not dense in $C(S)$. (vi) Let B be a maximal proper closed subalgebra of $C(X)$. We shall prove that B is not regular on X, and hence cannot contain A. The crucial fact about a maximal algebra B is the following. If K is any closed subset of X, then either $B|_K$ is dense in $C(K)$, or B contains every continuous function on X which vanishes on K. To prove this, let B' be the set of all f in $C(X)$ such that $f|_K$ is in the uniform closure of $B|_K$. It is clear that B' is a uniformly closed subalgebra of $C(X)$ which contains B. By the maximality of B, either $B' = C(X)$ or $B' = B$. If $B' = C(X)$, then every continuous function on K is a uniform limit of functions in $B|_K$. If $B' = B$, then every f in $C(X)$ for which $f|_K$ is in $B|_K$ lies in B. In particular, every f in $C(X)$ which vanishes on K is in the algebra B. With this fundamental fact established, we argue as follows. Let μ be a non-zero complex measure on X which annihilates B, and let K be the closed support of μ. Then $B|_K$ is not dense in $C(K)$. Let S be any proper closed subset of K. We shall prove that $A|_S$ is dense in $C(S)$. If $A|_S$ is not dense in $C(S)$, then every continuous function vanishing on S is in B. The restriction of μ to $K - S$ is then a measure which annihilates every continuous function on $K - S$ which vanishes at infinity. Hence, this restriction is 0 and μ is supported on S. This contradicts the definition of K. Now we refer to part (v) of the theorem and see that B cannot be regular on X, since we have produced a closed subset K of X such that $A|_K$ is not dense in $C(K)$, but $A|_S$ is dense in $C(S)$ for any proper closed subset S of K.

If we apply these results to the algebra A_α on the fiber $\mathfrak{M}_\alpha$, we have the following. The maximal ideal space of A_α is $\mathfrak{M}_\alpha$; the Šilov boundary for A_α is X_α; A_α is regular on X_α; A_α is contained in no maximal subalgebra of $C(X_\alpha)$; $\mathfrak{M}_\alpha$ is connected; X_α is totally disconnected. There is one further interesting property of A_α, which follows from the topological nature of X_α. In the section of this chapter which dealt with L^∞, we proved that X_α is a topological space with this property: if f is a continuous function on X_α, any value which f assumes is assumed on a non-empty open subset of X_α. By part (ii) of the last theorem, we then have the following. Let f be any function in the algebra A_α. Any value of f on the Šilov boundary X_α is assumed by f on a non-empty open subset of the maximal ideal space $\mathfrak{M}_\alpha$. We also see that, in some sense, A_α comes close to being normal on the space X_α. If K is any open-closed subset of X_α, there is a function f in A_α such that $f = 0$ on K and $|f| = 1$ on $X_\alpha - K$. If λ is in the range of f on X_α, then we have $f = \lambda$ on some non-empty open-closed subset K_λ of X_α. For each such λ of absolute value 1, the sets K and K_λ constitute a pair of disjoint closed sets for which we do have a function in A_α which vanishes on K and is 1 on K_λ. If ϕ is any complex homomorphism of A_α and S is a minimal support set for ϕ, we know that S cannot intersect both K and K_λ, and cannot intersect two distinct sets K_λ. Consequently, S has no interior in the space X_α, and (roughly speaking) every complex homomorphism of A_α is supported on a very "thin" subset of X_α.

Maximality

In this section we shall prove that H^∞ is a maximal weak-star closed subalgebra of L^∞, and that H^∞ is contained in no algebra which is maximal among the proper uniformly closed subalgebras of L^∞.

Theorem. *Let* B *be any (uniformly) closed subalgebra of* L^∞ *which contains* H^∞.

(i) *If* B $\neq H^\infty$, *then* B *contains* $\bar{z}$.

(ii) *If* B $\neq L^\infty$, *then there is some point α on the unit circle such that the restriction of* $\hat{B}$ *to the fiber* X_α *is a proper closed subalgebra of* $C(X_\alpha)$.

Proof. (i) Let ϕ_0 be the homomorphism of H^∞ given by evaluation at the origin. Either ϕ_0 extends to a complex homomorphism of the algebra B or it does not. Suppose that it does extend, i.e., that there is a complex homomorphism ψ of B such that $\psi(f) = f(0)$ for all f in H^∞. Through the Gelfand representation of L^∞, we have B represented as a uniformly closed subalgebra $\hat{B}$ of $C(X)$. Thus there is a positive measure μ on $X = \mathfrak{M}(L^\infty)$ such that

$$\psi(f) = \int_X \hat{f} d\mu, \quad f \text{ in } B.$$

This positive measure μ is, in particular, a representing measure for ϕ_0. By the Gleason-Whitney theorem, there is a unique representing measure m_0 for ϕ_0. Thus $\mu = m_0$ and

$$\psi(f) = \frac{1}{2\pi}\int_{-\pi}^{\pi} f(e^{i\theta})d\theta, \quad f \text{ in } B.$$

If f is any function in B, then for $n > 0$ we have

$$\frac{1}{2\pi}\int_{-\pi}^{\pi} e^{in\theta}f(e^{i\theta})d\theta = \psi(z^n f)$$
$$= \psi(z^n)\psi(f)$$
$$= 0 \cdot \psi(f)$$
$$= 0$$

from which we conclude that f is in H^∞. Thus, if $B \neq H^\infty$, the homomorphism ϕ_0 does not extend to B. Hence there is no complex homomorphism of B which sends z into 0, because ϕ_0 is the only complex homomorphism of H^∞ with this property. But then z lies in no proper ideal in the algebra B, which means that z is invertible in B, or that $1/z = \bar{z}$ is in B.

(ii) Suppose that B is a closed subalgebra of L^∞ which contains H^∞. If $|\alpha| = 1$, the function $f(z) = \frac{1}{2}(1 + \bar{\alpha}z)$ is in B and satisfies $\hat{f} = 1$ on X_α, $|\hat{f}| < 1$ on the remainder of X. As we have noted several times, this means that the algebra B_α, obtained by restricting $\hat{B}$ to X_α, is uniformly closed. Now suppose that $B_\alpha = C(X_\alpha)$ for each α on the circle; we shall prove that $B = L^\infty$. Let ϕ be any complex homomorphism of B, and let $\alpha = \phi(z)$. Then $|\alpha| = 1$, because, by part (i) of this theorem, $\bar{z}$ is in B and we must have $\phi(z)\overline{\phi(z)} = 1$. We again use the function f in B which is equal to 1 on X_α and of modulus less than 1 elsewhere on X. Since $\phi(z) = \alpha$, $\phi(f) = 1$, and any representing measure for ϕ must be supported on X_α. This means that ϕ is really a complex homomorphism of B_α, composed with the restriction map $B \to B_\alpha$. Since $B_\alpha = C(X_\alpha)$, ϕ must be evaluation at a point of X_α. We conclude that every complex homomorphism of B is evaluation (on $\hat{B}$) at a point of X, i.e., that X is the maximal ideal space of B. Now X is totally disconnected. The theorem of Šilov [82] states that $\hat{B}$ must contain the characteristic function of every open-closed subset of its maximal ideal space. Thus B contains the characteristic function of every measurable subset of the circle, so $B = L^\infty$.

Corollary. H^∞ *is a maximal weak-star closed subalgebra of* L^∞.

Proof. Let B be a weak-star closed subalgebra of L^∞ which contains H^∞. We prove that either $B = L^\infty$ or $B = H^\infty$. If $B \neq H^\infty$, part (i) of the theorem tells us that B contains $\bar{z}$. Since the trigonometric polynomials are weak-star dense in L^∞, we have $B = L^\infty$.

Corollary. *There is no algebra which contains* H^∞ *and is maximal among the proper uniformly closed subalgebras of* L^∞.

Proof. Let B be a proper closed subalgebra of L^∞ which contains H^∞. By part (ii) of the theorem, there is a point α on the unit circle such that B_α is a proper closed subalgebra of $C(X_\alpha)$. If B is maximal, then obviously B_α must be a maximal subalgebra of $C(X_\alpha)$. But B_α contains the algebra $A_\alpha = \hat{H}^\infty|_{X_\alpha}$, and we proved in the last section that A_α is contained in no maximal subalgebra of $C(X_\alpha)$. We conclude that B is not a maximal closed subalgebra of L^∞.

Interpolation

Let $\{z_k\}$ be a sequence of points in the open unit disc. We shall call $\{z_k\}$ an **interpolating sequence** if, for each bounded sequence of complex numbers $\{w_k\}$, there exists a function f in H^∞ such that $f(z_k) = w_k$ for every k. The main purpose of this section is to give a characterization of such sequences. We have chosen to incorporate this in our discussion of H^∞ as a Banach algebra because it is related to various questions about the maximal ideal space of H^∞. We might mention one example of this now. We have previously discussed the question of whether the open unit disc is dense in $\mathfrak{M}(H^\infty)$. We showed that this is equivalent to the following question: if $f_1, \ldots, f_n$ are functions in H^∞ such that

$$|f| + \cdots + |f_n| \geq \delta > 0$$

on the disc, do there exist functions $g_1, \ldots, g_n$ in H^∞ such that

$$f_1 g_1 + \cdots + f_n g_n = 1?$$

If $n = 2$ and $\{z_k\}$ is the sequence of zeros of f_1, then the function g_2 must interpolate the values $1/f_2(z_k)$ at the points z_k. If $\{z_k\}$ is an interpolating sequence and f_1 is simply the Blaschke product for that sequence, it is easy to see that g_1 and g_2 can be found. As we shall see later, the same result holds for the general n; that is, if one of the functions f_j is the Blaschke product for an interpolating sequence, then appropriate $g_1, \ldots, g_n$ can be found. Of course, what we are describing are very special cases of the density question in $\mathfrak{M}(H^\infty)$, but questions of this type led R. C. Buck to ask whether any interpolating sequences exist. L. Carleson, W. Hayman, and D. J. Newman worked independently on the interpolation problem, and each of them proved the existence of a great many interpolating sequences; e.g., every sequence which has a cluster point on the unit circle contains a subsequence which is an interpolating sequence. Both Carleson and Newman gave necessary and sufficient conditions for $\{z_k\}$ to be an interpolating sequence. Newman had two conditions on the sequence, whereas Carleson proved the stronger result, that one of these conditions is necessary and sufficient. Carleson's characterization of interpolating sequences is surprisingly simple, and, although it does not provide the solution of any deep problems about $\mathfrak{M}(H^\infty)$, it does shed some light on such problems.

Let us look at the interpolation problem. If $\{z_k\}$ is any sequence of points in the open unit disc, we consider the linear operator R which associates with each f in H^∞ its sequence of values on the points z_k:

$$Rf = (f(z_1), f(z_2), \ldots).$$

We wish to know under what condition R maps H^∞ onto ℓ^∞, the space of all bounded sequences of complex numbers. One obvious condition which is necessary is that the z_k be distinct. Another condition which is evidently necessary is that

$$\sum_{k=1}^\infty (1 - |z_k|) < \infty.$$

This condition is necessary in order to be able to find an f in H^∞ such that $f(z_k) = 0$ for $k \geq 2$ but $f(z_1) \neq 0$. If we employ a fundamental theorem on Banach spaces, we can deduce another necessary condition, which we shall subsequently show is sufficient. The space ℓ^∞ is a Banach space under the sup norm, and R is a norm-decreasing linear transformation of H^∞ into ℓ^∞. If R maps H^∞ onto ℓ^∞, then ℓ^∞ is isomorphic to the quotient space H^∞/I, where I is the ideal of all functions in H^∞ which vanish on the sequence $\{z_k\}$. If we equip H^∞/I with the quotient norm

$$||f + I|| = \inf_{g \in I} ||f + g||$$

then H^∞/I becomes a Banach space, and R induces a one-one norm-decreasing linear transformation of H^∞/I onto ℓ^∞. A corollary to the closed-graph theorem states that a one-one bounded linear transformation of a Banach space onto a Banach space necessarily has a bounded inverse. Thus there is a constant $M > 0$ such that, if $w = \{w_k\}$ is in ℓ^∞, there is an f in H^∞ with $Rf = w$ and

$$||f|| \leq M \sup |w_k|.$$

In other words, all sequences bounded by 1 can be interpolated in a uniformly bounded manner. In particular, for each k we can find a function f_k in H^∞ such that

$$f_k(z_j) = \delta_{jk}$$
$$||f_k|| \leq M.$$

If B_k denotes the Blaschke product whose zeros are the points z_j for $j \neq k$, then B_k divides f_k. Since $||f_k|| \leq M$ and $f_k(z_k) = 1$, we must have $|B_k(z_k)| \geq 1/M$. If we put $\delta = 1/M$, we have concluded that any interpolating sequence $\{z_k\}$ necessarily satisfies

(C) $$\prod_{j \neq k} \left| \frac{z_k - z_j}{1 - \bar{z}_j z_k} \right| \geq \delta > 0, \qquad k = 1, 2, 3, \ldots.$$

Carleson's theorem states that the condition (C) is also sufficient for $\{z_k\}$ to be an interpolating sequence. At the same time that Carleson proved this theorem, D. J. Newman independently showed that $\{z_k\}$ is an interpolating sequence if and only if it satisfies (C) together with the condition:

(N) $$\sum |f(z_k)|(1 - |z_k|) < \infty, \quad \text{for every } f \text{ in } H^1.$$

It follows that (C) implies (N), and the bulk of Carleson's work was devoted to establishing this implication. We are now going to prove the sufficiency of (C). The first step will be the proof of Newman's result. This is done by Banach space methods. Then, using the recent proof of Shapiro and Shields, we shall show that (C) implies (N).

Theorem. (Carleson; Newman). *Let $\{z_k\}$ be a sequence of points in the open unit disc. Then $\{z_k\}$ is an interpolating sequence if and only if conditions (C) and (N) are satisfied.*

Proof. Assume that we have a sequence of distinct points z_k. Suppose that $w = \{w_k\}$ is a bounded sequence of complex numbers. Certainly, we can find a function in H^∞ which interpolates w at any finite number of the points z_k. One way to do this is the following. Let

$$B_n(z) = \prod_{k=1}^n \frac{z - z_k}{1 - \bar{z}_k z}$$

$$B_{nk} = B_n(z) \cdot \frac{1 - \bar{z}_k z}{z - z_k}, \quad 1 \leq k \leq n$$

$$b_{nk} = B_{nk}(z_k).$$

Define

$$f_n(z) = \sum_{k=1}^n \frac{w_k}{b_{nk}} B_{nk}(z)$$

and we have $f_n(z_k) = w_k$, $1 \leq k < n$. In fact, the most general function in H^∞ which interpolates w at $z_1, \ldots, z_n$ is

$$f_n + B_n g, \quad g \text{ in } H^\infty.$$

If $\{z_k\}$ is an interpolating sequence, then all sequences w in the unit ball of ℓ^∞ can be interpolated in a uniformly bounded manner. This tells us that if we define

$$m_n(w) = \inf_{g \in H^\infty} ||f_n + B_n g||$$

we must have

$$\sup_n \sup_w m_n(w) < \infty, \quad ||w|| \leq 1.$$

But the converse also holds. If the above supremum is finite and w is an element of the unit ball of ℓ^∞, we can find a sequence of functions g_n in H^∞ such that

$$g_n(z_k) = w_k, \quad 1 \leq k \leq n$$

$$||g_n|| \leq M.$$

The sequence $\{g_n\}$ is a normal family and will then have a subsequence which converges uniformly on compact subsets of the disc to a function g in H^∞ with $||g|| \leq M$. It is clear that $g(z_k) = w_k$ for all k. We conclude that the finiteness of the supremum $\sup_n \sup_w m_n(w)$ is necessary and sufficient for $\{z_k\}$ to be an interpolating sequence.

Now

$$m_n(w) = \inf_{g \in H^\infty} ||f_n + B_n g||$$

$$= \inf_g ||F_n + g||_\infty$$

where $F_n = f_n/B_n$ and $||\cdots||_\infty$ denotes the essential sup norm on the unit circle. Here we have used the fact that $|B_n| = 1$ on the unit circle. Thus we see that $m_n(w)$ is the norm of the coset $F_n + H^\infty$ in the quotient space L^∞/H^∞. Now L^∞ is the conjugate space of the Banach space L^1. Also, H^∞ is the annihilator of the space H_0^1, consisting of the H^1 functions which vanish at the origin. Thus we may identify L^∞/H^∞ with the conjugate space of the Banach space H_0^1. In particular,

$$m_n(w) = \inf_{g \in H^\infty} ||F_n + g||_\infty$$

$$= \text{norm of } F_n \text{ as a linear functional on } H_0^1$$

$$= \sup_f \left| \frac{1}{2\pi} \int_{-\pi}^{\pi} F_n(e^{i\theta}) f(e^{i\theta}) e^{i\theta} d\theta \right|, \quad f \text{ in } H^1, \quad ||f||_1 \leq 1.$$

If f is in H^1,

$$\frac{1}{2\pi} \int_{-\pi}^{\pi} F_n f e^{i\theta} d\theta = \sum_{k=1}^{n} \frac{w_k}{b_{nk}} \frac{1}{2\pi} \int_{-\pi}^{\pi} \frac{B_{nk}}{B_n} f e^{i\theta} d\theta$$

$$= \sum_{k=1}^{n} \frac{w_k}{b_{nk}} \frac{1}{2\pi} \int_{-\pi}^{\pi} f(e^{i\theta}) (1 - \bar{z}_k e^{i\theta}) \frac{e^{i\theta}}{e^{i\theta} - z_k} d\theta$$

$$= \sum_{k=1}^{n} \frac{w_k}{b_{nk}} f(z_k)(1 - |z_k|^2).$$

Now we have

$$m_n(w) = \sup_f \left| \sum_{k=1}^{n} \frac{w_k}{b_{nk}} f(z_k)(1 - |z_k|^2) \right|, \quad f \text{ in } H^1, \quad ||f||_1 \leq 1.$$

Thus, if we constrain w to the unit ball in ℓ^∞,

$$\sup_w m_n(w) = \sup_f \sum_{k=1}^{n} \frac{1}{|b_{nk}|} |f(z_k)|(1 - |z_k|^2), \quad f \text{ in } H^1, \quad ||f||_1 \leq 1.$$

If we now let

$$M = \sup_f \sum_{k=1}^{\infty} |f(z_k)|(1 - |z_k|^2), \quad f \text{ in } H^1, \quad ||f||_1 \leq 1$$

and remember that $|b_{nk}| \leq 1$, we see that

$$M \leq \sup_n \sup_w m_n(w) \leq M \frac{1}{\inf |b_{nk}|}.$$

Condition (C) says that $\inf |b_{nk}| \geq \delta$, and thus we conclude that $\{z_k\}$ is an interpolating sequence if and only if (C) is satisfied and $M < \infty$. The theorem will now be proved if we show that $M < \infty$ if and only if

$$\sum |f(z_k)|(1 - |z_k|) < \infty$$

for every f in H^1. It is obviously immaterial whether we multiply $|f(z_k)|$ by $(1 - |z_k|)$ or by $(1 - |z_k|^2)$. If $M < \infty$, certainly

$$\sum |f(z_k)|(1 - |z_k|^2) < \infty$$

for each f in H^1. Conversely, if the sum is finite for each f, the rule
$$Tf = \{|f(z_k)|(1 - |z_k|^2)\}$$
defines a linear mapping of H^1 into the Banach space ℓ^1 of all absolutely summable sequences. The condition $M < \infty$ says simply that T is a bounded linear transformation from H^1 to ℓ^1. The closed-graph theorem asserts that T is bounded if its graph is closed. This means that if $f_n \to f$ in H^1 and the sequences Tf_n converge to the sequence $\lambda = \{\lambda_k\}$ in ℓ^1, then $Tf = \lambda$. Either the Poisson or Cauchy integral formula shows that if $f_n \to f$ in H^1, then $f_n(z) \to f(z)$ uniformly on compact subsets of the disc, so it is apparent that the graph of T is closed.

Actually we have proved a little more than we have stated in the last theorem. Before stating this extra information as a corollary, we want to switch from the space H^1 to the space H^2, in order to facilitate our later work. Consider the supremum
$$M = \sup_f \sum_{k=1}^{\infty} |f(z_k)|(1 - |z_k|^2), \quad f \text{ in } H^1, \quad \|f\|_1 \leq 1$$
which occurred in the proof. It is rather easy to see that
$$M = \sup_g \sum_{k=1}^{\infty} |g(z_k)|^2(1 - |z_k|^2), \quad g \text{ in } H^2, \quad \|g\|_2 \leq 1.$$
Since the square of any g in the unit ball of H^2 is a function in the unit ball of H^1, it is apparent that this second supremum does not exceed M. On the other hand, if f is any function in the unit ball of H^1, we can write $f = Bg^2$, where B is a Blaschke product and g is in the unit ball of H^2; also,
$$|f(z_k)| \leq |g(z_k)|^2.$$
This establishes the reverse inequality.

Corollary. *Let $\{z_k\}$ be a sequence of points in the unit disc which satisfies condition (C). Then $\{z_k\}$ is an interpolating sequence if and only if the supremum*
$$M = \sup_g \sum_{k=1}^{\infty} |g(z_k)|^2(1 - |z_k|^2), \quad g \text{ in } H^2, \quad \|g\|_2 \leq 1$$
is finite. When M is finite, every sequence in the unit ball of ℓ^∞ can be interpolated by a function of H^∞ whose norm does not exceed $\frac{1}{\delta} M$.

Now we set about the task of showing that condition (C) automatically guarantees the finiteness of M. We shall follow Shapiro and Shields in reducing this task to a weighted interpolation problem for H^2 functions. The reduction is contained in the following lemma.

Lemma 1. *Let $\{z_k\}$ be a sequence of points in the open unit disc. Suppose there exists a constant K such that if $\{\lambda_k\}$ is any square-summable sequence of complex numbers, then there is a g in H^2 satisfying*

(i) $\|g\|_2^2 \leq K \sum_{k=1}^{\infty} |\lambda_k|^2$.

(ii) $g(z_k)(1 - |z_k|^2)^{1/2} = \lambda_k$, $k = 1, 2, 3, \ldots$.

Then
$$\sum_{k=1}^{\infty} |g(z_k)|^2 (1 - |z_k|^2) \leq K \|g\|_2^2$$

for every **g** in H^2.

Proof. The proof will be virtually identical with the proof of the first theorem of this section, except that we replace H^∞ with H^2. We begin with a sequence $\lambda = \{\lambda_k\}$ satisfying $\sum |\lambda_k|^2 \leq 1$; that is, with a λ in the unit ball of the sequence space ℓ^2. As before, we consider the special functions f_n which solve the interpolation problem at the first n points. Because of the weights appearing in (ii) above, these functions are

$$f_n(z) = \sum_{k=1}^{n} \frac{\lambda_k}{b_{nk}} (1 - |z_k|^2)^{-1/2} B_{nk}(z).$$

As before, we define
$$m_n(\lambda) = \inf_{g \in H^2} \|f_n + B_n g\|_2$$
$$= \inf_{g \in H^2} \|F_n + g\|_2$$

where $F_n = f_n / B_n$. Thus $m_n(\lambda)$ is the norm of F_n as a linear functional on the space H_0^2. (Here, we are identifying the bounded linear functionals on L^2 with L^2 functions by omitting the complex conjugate which occurs in the definition of the inner product on L^2.) If we apply the Cauchy integral formula to the terms of $\int F_n f d\theta$, we obtain

$$m_n(\lambda) = \sup_f \left| \sum_{k=1}^{n} \frac{\lambda_k}{b_{nk}} f(z_k)(1 - |z_k|^2)^{1/2} \right|, \quad \|f\|_2 \leq 1.$$

Thus, if we constrain λ to the unit ball of ℓ^2,

$$\left[\sup_\lambda m_n(\lambda) \right]^2 \geq \sup_f \sum_{k=1}^{n} |f(z_k)|^2 (1 - |z_k|^2)$$

f in H^2, $\|f\|_2 \leq 1$. Therefore,

$$[\sup_n \sup_\lambda m_n(\lambda)]^2 \geq \sup_f \sum_{k=1}^{\infty} |f(z_k)|^2 (1 - |z_k|^2)$$

f in H^2, $\|f\|_2 \leq 1$. We are assuming that every λ in the unit ball of ℓ^2 can be interpolated as in (ii) by a function of H^2 whose square norm does not exceed K. Thus

$$[\sup_n \sup_\lambda m_n(\lambda)]^2 \leq K$$

and that completes the proof of the lemma.

Now we wish to show that any sequence $\{z_k\}$ which satisfies condition (C) also satisfies the hypothesis of Lemma 1. We shall do so in Lemma 4 below, but first we shall state two facts which, although elementary, are probably not obvious without some thought.

Lemma 2. *Let $\{a_{ij}\}$ be a double sequence of complex numbers which is self-adjoint, $a_{ji} = \overline{a_{ij}}$, and suppose there is a constant N such that*

$$\sum_{j=1}^{\infty} |a_{ij}| \leq N, \quad i = 1, 2, 3, \ldots.$$

Then for any square-summable sequence of complex numbers $\{\lambda_k\}$

$$\left| \sum_{i,j} a_{ij} \lambda_i \bar{\lambda}_j \right| \leq N \sum_{k=1}^{\infty} |\lambda_k|^2.$$

Proof. Suppose $A = [a_{ij}]$ is a self-adjoint $n \times n$ matrix such that $\sum_{j=1}^{n} |a_{ij}| \leq N$ for each i. Then

$$\sup_{\lambda} \left| \sum_{i,j} a_{ij} \lambda_i \bar{\lambda}_j \right|, \quad \sum |\lambda_k|^2 \leq 1$$

is the norm of A as a linear operator on complex Euclidean space C^n. This operator norm is, in turn, equal to the largest magnitude of any eigenvalue of A. If c is such an eigenvalue, then for some non-zero $n$tuple $(\lambda_1, \ldots, \lambda_n)$ we have

$$\sum_{i=1}^{n} a_{ij} \lambda_i = c \lambda_j, \quad j = 1, \ldots, n.$$

Thus

$$|c||\lambda_j| \leq \sum_{i=1}^{n} |a_{ij}||\lambda_i|.$$

If we sum these inequalities on j and divide by $\sum |\lambda_j|$, we see that $|c| \leq N$. This proves the lemma in the finite case. In fact, if we replace a_{ij} by $|a_{ij}|$, the conditions are not changed, so

$$\sum_{i=1}^{n} \sum_{j=1}^{n} |a_{ij}||\lambda_i||\lambda_j| \leq N \sum_{k=1}^{n} |\lambda_k|^2$$

for each n. The lemma is now obvious.

Lemma 3. *If $\{z_k\}$ is a sequence which satisfies condition (C), then*

$$\sum_{j=1}^{\infty} \frac{(1 - |z_j|^2)(1 - |z_k|^2)}{|1 - \bar{z}_j z_k|^2} \leq 1 - 2 \log \delta, \quad k = 1, 2, 3, \ldots.$$

Proof. It is easy to verify that

$$\left| \frac{z_k - z_j}{1 - \bar{z}_j z_k} \right|^2 = 1 - \frac{(1 - |z_j|^2)(1 - |z_k|^2)}{|1 - \bar{z}_j z_k|^2}.$$

Condition (C) tells us that

$$\prod_{j \neq k} \left| \frac{z_k - z_j}{1 - \bar{z}_j z_k} \right|^2 \geq \delta^2$$

or that

$$-\sum_{j \neq k} \log \left| \frac{z_k - z_j}{1 - \bar{z}_j z_k} \right|^2 \leq -2 \log \delta.$$

If one now uses the identity at the beginning of the proof and the inequality

$$x \leq -\log(1 - x)$$

the conclusion follows.

Lemma 4. *If the sequence $\{z_k\}$ satisfies condition (C), then for any square-summable sequence $\{\lambda_k\}$ there is a function g in H^2 such that*

(i) $\|g\|_2^2 \leq \dfrac{2}{\delta^4} (1 - 2 \log \delta) \sum_k |\lambda_k|^2$.

(ii) $g(z_k)(1 - |z_k|^2)^{1/2} = \lambda_k, \quad k = 1, 2, 3, \ldots$.

Proof. Once again, we begin with a square-summable sequence $\{\lambda_k\}$ and solve the weighted interpolation problem at the first n points; however, we now define

$$g_{nk}(z) = \left[\frac{B_n(z)}{z - z_k}\right]^2 (1 - |z_k|^2)^{3/2}$$

$$f_n(z) = \sum_{k=1}^{n} \frac{\lambda_k}{b_{nk}^2} g_{nk}(z).$$

It is easy to see that

$$f(z_k)(1 - |z_k|^2)^{1/2} = \lambda_k, \quad 1 \leq k \leq n.$$

Now we shall obtain a uniform bound on the norms of the f_n. First

$$\|f_n\|_2^2 = (f_n, f_n) = \sum_{i,j=1}^{n} \frac{\lambda_i \bar{\lambda}_j}{b_{ni}^2 b_{nj}^2} (g_{ni}, g_{nj})$$

where (f, g) denotes the usual inner product on L^2 of the circle. Now

$$(g_{ni}, g_{nj}) = (1 - |z_i|^2)^{3/2}(1 - |z_j|^2)^{3/2}([z - z_i]^{-2}, [z - z_j]^{-2})$$

since B_n is of modulus 1 on the circle. If one uses the Cauchy integral formula, it is relatively easy to calculate the last inner product and obtain

$$(g_{ni}, g_{nj}) = (1 - |z_i|^2)^{3/2}(1 - |z_j|^2)^{3/2} \frac{1 + z_i \bar{z}_j}{(1 - z_i \bar{z}_j)^3}$$

Since $(1 - |z_i|^2)^{1/2}(1 - |z_j|^2)^{1/2} \leq |1 - \bar{z}_i z_j|$, we see from Lemma 3 that

$$\sum_j |(g_{ni}, g_{nj})| \leq 2(1 - 2 \log \delta).$$

Then, by Lemma 2, we have

$$\|f_n\|_2^2 \leq \frac{2}{\delta^4}(1 - 2 \log \delta) \sum_k |\lambda_k|^2.$$

Since the functions f_n all lie in a fixed ball in H^2, some subsequence of $\{f_n\}$ will converge weakly to a function g in that ball. This weak convergence guarantees (at least) pointwise convergence on the unit disc, so g

will have properties (i) and (ii) in the statement of the theorem. That completes the proof.

If we now combine Lemmas 1 and 4 with the corollary to the first theorem of this section, we have the result we have been seeking.

Theorem. *Let $\{z_k\}$ be a sequence of points in the open unit disc. A necessary and sufficient condition that $\{z_k\}$ be an interpolating sequence is that there exist a positive number δ such that*

$$\prod_{j \neq k} \left| \frac{z_k - z_j}{1 - \bar{z}_j z_k} \right| \geq \delta, \qquad k = 1, 2, 3, \ldots.$$

If this condition is satisfied, then for any bounded sequence $\{w_k\}$ there is an f in H^∞ such that

(i) $f(z_k) = w_k, \qquad k = 1, 2, 3, \ldots,$

(ii) $\|f\| \leq \dfrac{2}{\delta^5} (1 - 2 \log \delta) \sup_k |w_k|.$

Before presenting some corollaries, we should like to make one remark about the proof of Lemma 4. The functions f_n which were chosen there to interpolate at the first n points are not the ones of minimal L^2 norm. This shrewd choice of the f_n by Shapiro and Shields accounts in part for the relatively short proof of Lemma 4, which they discovered.

Corollary (Hayman; Newman). *Suppose $\{z_k\}$ is a sequence of points in the open unit disc which approaches the boundary exponentially, i.e.,*

$$\frac{1 - |z_n|}{1 - |z_{n-1}|} < c < 1.$$

Then $\{z_k\}$ is an interpolating sequence.

Proof. Since, for points α, β in the unit disc,

$$\left| \frac{\alpha - \beta}{1 - \bar{\alpha}\beta} \right| \geq \frac{|\alpha| - |\beta|}{1 - |\alpha||\beta|}$$

we see that

$$\prod_{j \neq k} \left| \frac{z_k - z_j}{1 - \bar{z}_j z_k} \right| \geq \prod_{j < k} \frac{|z_k| - |z_j|}{1 - |z_j||z_k|} \prod_{j < k} \frac{|z_j| - |z_k|}{1 - |z_j||z_k|}.$$

When $j > k$, we have

$$1 - |z_j| \leq c^{j-k}(1 - |z_k|)$$

and thus

$$|z_j| - |z_k| \geq (1 - c^{j-k})(1 - |z_k|).$$

On the other hand,

$$1 - |z_j||z_k| \leq (1 + c^{j-k})(1 - |z_k|).$$

Thus
$$\prod_{j>k} \geq \prod_{n=1}^{\infty} \frac{1-c^n}{1+c^n}.$$

If $j < k$,
$$1 - |z_k| \leq c^{k-j}(1 - |z_j|)$$
$$|z_k| - |z_j| \geq (1 - c^{k-j})(1 - |z_j|)$$
$$1 - |z_k||z_j| \leq (1 + c^{k-j})(1 - |z_j|).$$

Thus
$$\prod_{j<k} \geq \prod_{n=1}^{\infty} \frac{1-c^n}{1+c^n}$$

and it is clear that $\{z_k\}$ satisfies condition (C).

Corollary. *Any $\{z_k\}$ such that $\overline{\lim} |z_k| = 1$ contains a subsequence which is an interpolating sequence.*

Proof. This is clear from the previous corollary.

Corollary (Hayman; Newman). *If $\{z_k\}$ is an increasing sequence of points on the positive axis, then $\{z_k\}$ is an interpolating sequence if and only if*
$$\frac{1 - z_n}{1 - z_{n-1}} < c < 1.$$

Proof. We have shown that interpolation is possible if the z_k tend to the boundary exponentially. Conversely, if interpolation is possible, there is a positive δ such that
$$\delta \leq \frac{z_k - z_{k-1}}{1 - z_k z_{k-1}} \leq \frac{z_k - z_{k-1}}{1 - z_{k-1}} = 1 - \frac{1 - z_k}{1 - z_{k-1}}.$$

These last corollaries show us that if $|z_k|$ tends to 1 in a sufficiently rapid way, then $\{z_k\}$ is an interpolating sequence; however, one should not get the idea that any such growth condition is necessary for interpolation, other than the obvious one:
$$\sum_{k=1}^{\infty} (1 - |z_k|) < \infty.$$

Three years before the solution of the H^∞ interpolation problem, A. G. Naftalevitch showed the following. If $\{z_k\}$ is any sequence of non-zero numbers which satisfies this last summability condition, there exists a sequence $\{\tilde{z}\}$ which satisfies condition (C) and for which
$$|\tilde{z}_k| = |z_k|, \quad k = 1, 2, 3, \ldots.$$

There are different ways to formulate necessary and sufficient conditions for $\{z_k\}$ to be an interpolating sequence. The condition (C) is certainly the simplest, but there are others which are instructive. One of these is this: every sequence of 0's and 1's can be interpolated by some H^∞

function. Hayman has given a direct proof that if every such idempotent sequence can be interpolated, then the sequence $\{z_k\}$ satisfies condition (C). When combined with Carleson's result, this shows the equivalence of interpolation of all bounded sequences and interpolation of idempotents. Actually, this equivalence is a consequence of a more general result of Bade and Curtis on Banach algebras.

Theorem. *Let* B *be a commutative Banach algebra with identity and let* $S = \{p_k\}$ *be a sequence of distinct points in the maximal ideal space of* B. *The following are equivalent.*

(i) $\hat{B}|_S = \ell^\infty$.

(ii) $\hat{B}|_S$ *contains every idempotent in* ℓ^∞.

(iii) S *is discrete in its relative topology as a subset of* $\mathfrak{M}(B)$; *its closure* $\bar{S}$ *in* $\mathfrak{M}(B)$ *is homeomorphic to the Čech compactification of* S; $\bar{S}$ *is a hull in* $\mathfrak{M}(B)$.

Proof. Obviously, (i) implies (ii). Now we shall prove that (ii) implies (i). Here we shall use the result of Badé and Curtis [Corollary 3.5, page 858, *Amer. Jour. Math.*, October, 1960]. It states the following. Suppose that A is a complex linear subalgebra of ℓ^∞ which contains every idempotent in ℓ^∞. If there is any norm on A under which it is a Banach algebra, then $A = \ell^\infty$. In the case at hand, we apply this theorem to $A = \hat{B}|_S$. All that we need demonstrate is that this A is a Banach algebra with some suitable norm. But this is clear. Let I be the (closed) ideal consisting of those elements x in B such that $\hat{x} = 0$ on S. Then

$$(x + I) \leftrightarrow \hat{x}|_S$$

is an isomorphism between the quotient algebra B/I and the algebra $\hat{B}|_S$. The standard quotient norm on B/I is a Banach algebra norm, and that completes the argument that (ii) implies (i). Now assume that (i) holds. We shall prove (iii). Since every bounded function on the sequence S is the restriction to S of a continuous function $\hat{x}$, it is clear that each p_k is isolated from the points p_j, $j \neq k$. In other words, S is discrete as a topological subspace of $\mathfrak{M}(B)$. Consider the closure $\bar{S}$ of S in $\mathfrak{M}(B)$. Each bounded function on S has a continuous extension to $\bar{S}$, by (i). The functions $\hat{x}$, x in B, separate the points of $\bar{S}$, since they separate the points of $\mathfrak{M}(B)$. Since S is dense in $\bar{S}$, it follows that $\bar{S}$ is the Čech compactification of S, i.e., the smallest compact Hausdorff space which contains S and has the property that every bounded (continuous) function on S has a continuous extension to the containing space. The *hull* of S is the set of all points p in $\mathfrak{M}(B)$ such that $\hat{x}(p) = 0$ for every $\hat{x}$ which vanishes on S, i.e., for every x in the ideal I. For any p in the hull of S,

$$\hat{x}|_S \to \hat{x}(p)$$

defines a complex homomorphism of the algebra $\hat{B}|_S$. But $\hat{B}|_S = \ell^\infty$, which is isomorphic to the algebra of all continuous functions on $\bar{S}$. The above homomorphism is, therefore, evaluation at a point of $\overline{S}$, that is, p is in $\overline{S}$. We conclude that $\overline{S} = $ hull (S). That completes the proof that (i) implies (iii). Assume now that (iii) holds. We prove (ii). As we noted before, $\hat{B}|_S$ is a commutative Banach algebra, using the norm inherited from B/I. It is easy to verify that every complex homomorphism of $B|_S$ is evaluation at a point of the hull of S. Since hull $(S) = \overline{S}$ and $\overline{S}$ is the Čech compactification of the discrete countable space S, the maximal ideal space of $\hat{B}|_S$ is the totally disconnected space $\overline{S}$. By the theorem of Šilov (82), $\hat{B}|_S$ must contain every idempotent continuous function on $\overline{S}$. Hence, we have (ii).

If we apply this theorem to H^∞ and combine with our earlier results, we see the following. If $S = \{z_k\}$ is a sequence of distinct points in the open unit disc, these are equivalent: (i) S is an interpolating sequence; (ii) S satisfies condition (C); (iii) primitive idempotent sequences (i.e., sequences of one 1 and the remainder 0's) can be interpolated in a bounded way; (iv) every idempotent sequence can be interpolated; (v) the closure of the sequence S in the maximal ideal space of H^∞ is homeomorphic to the Čech compactification of the integers, and if B is the Blaschke product with zeros $\{z_k\}$, then every zero of $\hat{B}$ on $\mathfrak{M}(H^\infty)$ is in the closure of S.

Corollary. *Let* B *be a Blaschke product whose zeros are an interpolating sequence, and let* $f_1, \ldots, f_n$ *be functions in* H^∞. *The following are equivalent.*

(i) *There are functions* g, $g_1, \ldots, g_n$ *in* H^∞ *such that*

$$gB + \sum_j g_j f_j = 1.$$

(ii) *There is a* $\delta > 0$ *such that*

$$|B| + |f_1| + \cdots + |f_n| \geq \delta$$

on the unit disc.

(iii) *There is a* $\delta > 0$ *such that*

$$|f_1| + \cdots + |f_n| \geq \delta$$

on the sequence of zeros of B.

NOTES

For a discussion of the maximal ideal space of a commutative Banach algebra, see the papers of Gelfand [32], Gelfand-Raikov-Šilov [33], or the book by Loomis [54]. The results on $\mathfrak{M}(H^\infty)$ which are found in the second, third, and fourth sections of this chapter are contained in the paper of I. J. Schark [80]. The basic results on L^∞ as a Banach algebra are well-known; e.g., see Dunford-Schwartz [25]. The identification of the Šilov boundary for H^∞ is also from the paper of I. J. Schark [80]. The uniqueness of representing measures for points in the open unit disc was proved in the paper of Gleason and Whitney [35], together with various generalizations. It was first pointed out by Gleason that there are no non-trivial Dirichlet algebras on a totally disconnected space. The proof given here was shown to me by H. S. Bear. The regularity of the algebra A_α on the fiber X_α was proved in the paper of Hoffman and Singer [49], together with these consequences: A_α lies in no maximal subalgebra of $C(X_\alpha)$, and H^∞ lies in no maximal subalgebra of L^∞. The connectedness of the fibers $\mathfrak{M}_\alpha$ and the space $\mathfrak{M}\text{-}\Delta$ was proved in the paper of Hoffman [48]. The various consequences of the fact that the algebra A is regular on the space X do not depend upon the fact that A is uniformly closed, but only on the fact that A is a commutative Banach algebra under some norm. The basic material on the interpolation problem in H^∞ is contained in the papers of Carleson [18], Newman [65], and Hayman [Ann. de l'Institute Fourier, XIII (1958)]. The proofs we have used here are in the paper of Shapiro and Shields [81]. Their paper contains a nice discussion of this and other interpolation problems, and a reasonable bibliography on these problems. While reading the page proofs for this chapter, I received a manuscript from Lennart Carleson, and it appears that he has proved that the open unit disc is dense in $\mathfrak{M}(H^\infty)$.

EXERCISES

1. Prove that the complement of the open unit disc in $\mathfrak{M}(H^\infty)$ is the maximal ideal space of the closed subalgebra of L^∞ which is generated by H^∞ and $\bar{z}$.

2. Let A be a uniformly closed subalgebra of $C(X)$. Let ϕ be a complex homomorphism of A which is not evaluation at a point of X.

(a) Among all the support sets for ϕ prove there is a minimal one (not necessarily a minimum one).

(b) Prove that any minimal support set for ϕ is a perfect set. (Hint: If x_0 is an isolated point of S_ϕ, the closed support of a representing measure m_ϕ for ϕ, choose $f \in A$ such that $f(x_0) = 0$ and $\phi(f) = 1$, and look at fdm_ϕ.)

3. Consider the algebra A_α, obtained by restricting $\hat{H}^\infty$ to the fiber $\mathfrak{M}_\alpha$ in $\mathfrak{M}(H^\infty)$. Prove the following:

(a) A_α contains a non-constant real-valued function.

(b) Every point on the Šilov boundary X_α is in the closure of $\mathfrak{M}_\alpha - X_\alpha$.

(c) If K is any open-closed subset of X_α, then the restriction of A_α to K is not dense in $C(K)$.

(d) If K is a closed subset of X_α, then the restriction of A_α to K is $C(K)$, if and only if every measure on X_α which is orthogonal to A_α has total variation 0 on K.

4. With the notation of Exercise 3, if ϕ is a complex homomorphism of A_α, let S_ϕ be the closed support of the (unique) representing measure for ϕ. Prove the following:

(a) If $\phi \in \mathfrak{M}_\alpha - X_\alpha$, then S_ϕ is a perfect subset of X_α which has no interior in X_α.

(b) For any $\phi \in \mathfrak{M}_\alpha$ the support set S_ϕ is an intersection of "peak" sets for H^∞, i.e., subsets of $\mathfrak{M}(H^\infty)$ of the form $\{\hat{f} = 1\}$, where $f \in H^\infty$ and $||f|| = 1$.

(c) If $\phi \in \mathfrak{M}_\alpha - X_\alpha$, there exists a $\psi \in \mathfrak{M}_\alpha - X_\alpha$ such that S_ψ is a proper closed subset of S_ϕ.

(d) If K is an open-closed subset of X_α, then the restriction of A_α to K is not uniformly closed, unless K is empty or $K = X_\alpha$.

5. Let $\{z_k\}$ be a sequence of points in the open unit disc. Prove that $S = \{z_k\}$ is an interpolation sequence if and only if the following is true: If S_1 and S_2 are any two disjoint subsequences of S with corresponding Blaschke products B_1 and B_2, then B_1 and B_2 lie in no proper ideal of H^∞.

Suppose $\{z_k\}$ and $\{\tilde{z}_k\}$ are disjoint interpolating sequences, with corresponding Blaschke products B and $\tilde{B}$. Prove that the union of the two sequences is an interpolating sequence if and only if B and $\tilde{B}$ lie in no proper ideal of H^∞.

6. Prove the equivalence of the three statements in the final Corollary of this Chapter. Give an example of a Blaschke product for which these statements are equivalent, but whose zeros do not form an interpolating sequence.

7. Let A be the uniform closure of the polynomials on the unit disc. If $\{z_k\}$ is a sequence of distinct points in the open unit disc, show that the following two statements are equivalent.

(i) If g is any continuous function on the closed unit disc, there exists $f \in A$ such that $f(z_k) = g(z_k)$, $k = 1, 2, 3, \ldots$.

(ii) $\{z_n\}$ is an interpolating sequence for H^∞, and the set of accumulation points of $\{z_k\}$ on the unit circle has Lebesgue measure zero.

BIBLIOGRAPHY

1. Ahiezer, N. I., *Theory of Approximation*, Ungar Publishing Co., New York, 1956.
2. Ahlfors, L., *Complex Analysis*, McGraw-Hill, New York, 1953.
3. Akutowitz, E. J., *A qualitative characterization of Blaschke products in a half-plane*, Amer. Jour. Math., vol. 78, 1956.
4. Arens, R., *The boundary integral of* $\log |\phi|$ *for generalized analytic functions*, Trans. A.M.S., vol. 86, 1957.
5. Arens, R. and Singer, I. M., *Function values as boundary integrals*, Proc. A.M.S., vol. 5, 1954.
6. ———, *Generalized analytic functions*, Trans. A.M.S., vol. 81, 1956.
7. Banach, S., *Théorie des Opérations Linéaires*, Warsaw, 1932.
8. Beurling, A., *On two problems concerning linear transformations in Hilbert space*, Acta Math., vol. 81, 1949.
9. Bieberbach, L., *Lehrbuch der Functionentheorie* (two volumes), Leipzig, 1927.
10. Bishop, E., *Boundary measures of analytic differentials*, Duke Math. Jour., vol. 27, 1960.
11. ———, *A generalized Rudin-Carleson theorem* (to appear in Proc. A.M.S.).
12. Bochner, S., *Vorlesungen über Fouriersche Integrale*, Leipzig, 1932.
13. ———, *Boundary values of analytic functions in several variables and of almost periodic functions*, Ann. of Math., vol. 45, 1944.
14. ———, *Generalized conjugate and analytic functions without expansions*, Proc. Nat. Acad. Sci., vol. 45, 1959.
15. ———, *Additive set functions on groups*, Ann. of Math., vol. 40, 1939.
16. Caratheodory, C., *Über die Fourierschen Koeffizienten monotoner Functionen*, Sitzber preuss Akad. Wiss. Berlin, vol. 30, 1920.
17. Carleson, L., *Representations of continuous functions*, Math. Zeit., vol. 66, 1957.
18. ———, *On bounded analytic functions and closure problems*, Ark. for Mat., 1952.
19. Carlson, F., *Quelques inequalites concernant les fonctions analytiques*, Arkiv Mat. Astron. Fysik, vol. 11, 1943.

20. Cohen, P., *A note on constructive methods in Banach algebras*, Proc. A.M.S., vol. 12, 1961.
21. Courant, R., *Dirichlet's Principle, Conformal Mapping, and Minimal Surfaces*, Interscience, New York, 1950.
22. deLeeuw, K., *The isometries of H^1* (mimeographed note, Stanford, 1960).
23. deLeeuw, K. and Rudin, W., *Extreme points and extremum problems in H^1*, Pac. Jour. Math., vol. 8, 1958.
24. deLeeuw, K., Rudin, W., and Wermer, J., *The isometries of some function spaces*, Proc. A.M.S., vol. 11, 1960.
25. Dunford, N. and Schwartz, J., *Linear Operators* (Part I), Interscience, New York, 1958.
26. Evans, G. C., *The Logarithmic Potential*, A.M.S. Colloq. Publ., vol. VI, 1927.
27. Fatou, P., *Séries trigonométriques et séries de Taylor*, Acta Math., vol. 30, 1906.
28. Fejér, L., *Untersuchungen über Fouriersche Reihen*, Mat. Ann., vol. 58, 1904.
29. Gabriel, R. M., *Concerning the zeros of a function regular in a half-plane*, Jour. London Math. Soc., vol. 4, 1929.
30. ———, *An improved result concerning the zeros of a function regular in a half-plane*, Jour. Lond. Math. Soc., vol. 4, 1929.
31. ———, *An inequality concerning the integrals of positive subharmonic functions along certain curves*, Jour. London Math. Soc., vol. 5, 1930.
32. Gelfand, I. M., *Normierte Ringe*, Mat. Sbornik, 1941.
33. Gelfand, I., Raikov, D., and Šilov, G., *Commutative normed rings*, Amer. Math. Soc. Translations Series 2, vol. 5, 1957.
34. Gleason, A., *Function algebras*, Seminars on Analytic Functions, Institute for Advanced Study, Princeton, 1957.
35. Gleason, A. and Whitney, H., *The extension of linear functionals defined on H^∞*, Pacific Jour. Math., to appear.
36. Gross, W., *Zur Poissonschen Summierung*, Sitzber. Akad. Wiss. Wien, Math. Naturw. Klasse., vol. 124, 1915.
37. Halmos, P. R., *Introduction to Hilbert Space*, Chelsea, New York, 1951.
38. ———, *Measure Theory*, Van Nostrand, New York, 1950.
39. ———, *Shifts on Hilbert spaces*, Journal für Reine und Angewende Mat., 1961.
40. Hardy, G. H., *Theorems relating to summability and convergence of slowly oscillating series*, Proc. London Math. Soc., vol. 8, 1910.
41. Hardy, G. H. and Littlewood, J. E., *Some new properties of Fourier constants*, Mat. Ann., vol. 97, 1926.
42. Helson, H., *On a theorem of F. and M. Riesz*, Colloq. Math., vol. 3, 1955.
43. Helson, H. and Lowdenslager, D., *Prediction theory and Fourier series in several variables*, Acta Math., vol. 99, 1958.

44. ———, *Invariant subspaces*, Proc. of 1960 Jerusalem Conf. in Anal.

45. Herglotz, G., *Über Potenzreihen mit positivem, reellem Teil im Einheitskreis*, Ber. Verhandl. sächs. Akad. Wiss. Leipzig, Math-phys. Klasse, vol. 63, 1911.

46. Hille, E. and Tamarkin, J. D., *On the absolute integrability of Fourier transforms*, Fund. Math., vol. 25, 1935.

47. Hoffman, K., *Boundary behavior of generalized analytic functions*, Trans. A.M.S., vol. 87, 1958.

48. ———, *A note on the paper of I. J. Schark*, Jour. Math. and Mech., 1961.

49. Hoffman, K. and Singer, I. M., *Maximal algebras of continuous functions*, Acta Math., vol. 103, 1960.

50. Kakutani, S., *Rings of analytic functions*, Lectures on functions of a complex variable, Ann Arbor, 1955.

51. Karhunen, K., *Über die Struktur stationärer zufälliger Funktionen*, Ark. Mat., vol. 1, 1950.

52. Krein, M. and Milman, D., *On extreme points of regular convex sets*, Stud. Math., vol. 9, 1940.

53. Lax, P., *Translation invariant subspaces*, Acta Math., vol. 101, 1959.

Littlewood, J. E., and Hardy, G. H. (*see* Hardy and Littlewood).

54. Loomis, L., *An Introduction to Abstract Harmonic Analysis*, Van Nostrand, New York, 1953.

55. Loomis, L. and Widder, D., *The Poisson integral representation of functions which are positive and harmonic in a half-plane*, Duke Math. Jour., vol. 9, 1942.

Lowdenslager, D., and Helson, H. (*see* Helson and Lowdenslager).

56. Lusin, N. and Privaloff, J., *Sur l'unicité et la multiplicité des fonctions analytiques*, Ann. Sci. Ec. Norm. Sup., vol. 42, 1925.

57. Mackey, G., *The Laplace transform for locally compact abelian groups*, Proc. Nat. Acad. Sci., vol. 34, 1948.

58. Masani, P., four notes in Compt. rend., vol. 249, 1959.

59. Masani, P. and Wiener, N., *The prediction theory of multivariate stochastic processes*, I and II, Acta Math., vols. 97 and 98, 1957, 1958.

60. Mergelyan, S., *On the representation of functions by series of polynomials on closed sets*, A.M.S. Translation No. 85.

61. ———, *Uniform approximations to functions of a complex variable*, A.M.S. Translation No. 101.

62. Müntz, Ch., *Über den Approximationsatz von Weierstrass*, H. A. Festschrift, Berlin, 1914.

63. Nagasawa, M., *Isomorphisms between commutative Banach algebras with an application to rings of analytic functions*, Kokai Math. Sem. Rep., vol. 11, 1959.

64. Nevanlinna, R., *Eindeutige Analytische Funktionen*, Berlin, 1936.

65. Newman, D. J., *Interpolation in H^∞*, Trans. A.M.S., vol. 92, 1959.
66. ———, *The non-existence of projections from L^1 to H^1*, Proc. A.M.S., vol. 12, 1961.
67. Paley, R. E. A. C. and Wiener, N., *Notes on the theory and application of Fourier transforms*, I–II. Trans. A.M.S., vol. 35, 1933.
68. ———, *Fourier transforms in the complex domain*, Amer. Math. Soc. Colloq. Publ., Vol. XIX, New York, 1934.
69. Potapov, V. L., *On matrix functions holomorphic and bounded in the unit circle*, Doklady Akad. Nauk S.S.S.R., vol. 72, 1950.
70. Privaloff, I., *Randeigenschaften Analytischer Funktionen*, Deutscher Verlag der Wiss., Berlin, 1956.

Privaloff, I. and Lusin, N. (*see* Lusin and Privaloff).

71. Riesz, F., *Über die Randwerte einer analytischen Funktion*, Mat. Zeit., vol. 18, 1922.
72. Riesz, F. and M., *Über die Randwerte einer analytischen Funktion*, Quatrieme Congrès des Math. Scand., 1916.
73. Riesz, F. and Sz.-Nagy, B., *Leçons D'Analyse Fonctionelle*, Paris-Budapest, 1955.

Riesz, M. (*see* Riesz, F).

74. Rudin, W., *Boundary values of continuous analytic functions*, Proc. A.M.S., vol. 7, 1956.
75. ———, *The closed ideals in an algebra of continuous functions*, Can. Jour. Math., vol. 9, 1957.
76. ———, *Fourier Analysis on Groups*, Interscience (in print).
77. ———, *Projections on invariant subspaces*, Univ. of Wisconsin MRC Technical Report 222, 1961.
78. Runge, C., *Zur Theorie der eindeutigen analytischen Functionen*, Acta Math., vol. 6, 1885.
79. Saks, S., *Theory of the Integral*, Warsaw, 1937.
80. Schark, I. J., *The maximal ideals in an algebra of bounded analytic functions*, Jour. Math. and Mech., to appear 1961.

Shields, A. L., and Shapiro, H. S. (*see* Shapiro and Shields).

81. Shapiro, H. S. and Shields, A. L., *On some interpolation problems for analytic functions*, to appear in Amer. Jour. Math.
82. Šilov, G. E., *On the decomposition of a commutative normed ring into a direct sum of ideals*, Mat. Sbornik, vol. 32, 1953.

Singer, I. M. and Arens, R. (*see* Arens and Singer).

Singer, I. M. and Hoffman, K. (*see* Hoffman and Singer).

83. Steinhaus, H., *Sur quelques propriétés des séries trigonométric et de celles de Fourier*, Rozprawy Akademji Umiejetnosci, Cracow, vol. 56, 1925.

84. Stone, M., *Linear Transformations in Hilbert Space*, A.M.S. Colloq. Publ. XV, 1932.

85. Szasz, O., *Über die Approximation stetiger Functionen durch lineare Aggregate von Potenzen*, Math. Ann., vol. 77, 1916.

86. Szegö, G., *Beiträge zur Theorie der toeplitzen Formen (Ersten Mitteilung)*, Math. Zeit, vol. 6, 1920.

Tamarkin, J. D., and Hille, E. (*see* Hille and Tamarkin).

87. Titchmarsh, E. C., *Theory of Functions*, Oxford University Press, 1932.

88. ———, *Introduction to the Theory of Fourier Integrals*, Clarendon Press, Oxford, 1937.

89. Walsh, J. L., *Interpolation and approximation by rational functions in the complex domain*, Amer. Math. Soc. Colloq. Publ. Vol. XX, New York, 1935.

90. Wermer, J., *On algebras of continuous functions*, Proc. A.M.S., vol. 4, 1953.

91. ———, *Subalgebras of the algebra of all complex-valued continuous functions on the circle*, Amer. Jour. Math., vol. 78, 1954.

92. ———, *Banach algebras and analytic functions*, Academic Press (to appear).

93. ———, *Dirichlet algebras*, Duke Jour. of Math., vol. 27, 1960.

Wermer, J., deLeeuw, K., and Rudin, W. (*see* deLeeuw, Rudin, Wermer).

Whitney, H. and Gleason, A. (*see* Gleason and Whitney).

Widder, D. and Loomis, L. (*see* Loomis and Widder).

94. Wiener, N., *Tauberian theorems*, Ann. Math., vol. 33, 1932.

95. ———, *The Fourier Integral and Certain of its Applications*, Cambridge University Press, 1933.

Wiener, N. and Masani, P. (*see* Masani and Wiener).

Wiener, N. and Paley, R. E. A. C., *see* Paley and Wiener.

96. Young, W. H., *On the condition that a trigonometrical series should have a certain form*, Proc. Lond. Roy. Soc. (A)88, 1913.

97. Young, G. C. and W. H., *On the theorem of Riesz-Fischer*, Quart. Journ. Math., vol. 44, 1913.

98. Zygmund, A., *Trigonometric Series*, 2d Ed., Cambridge University Press, 1959.

INDEX

A, 42, 54, 77
A_0, 44, 55
Abel summability, 32, 33, 34
Absolute continuity, 4
  convergence, 71, 96, 97
Absolutely continuous function, 71
  measure, 21, 44, 46, 47, 55
Almost everywhere (a.e.), 3
Analytic function, 27, 114
  measure, 39, 47, 51
Approximate identity, 17, 22, 32, 35
Automorphism, 143, 157

Baire function, 2
  measure, 1
  set, 1
Banach algebra, 90, 205
  H^∞ as one, 158
  L^∞ as one, 169
Banach space, 5
Bessel's inequality, 11
Blaschke product, 66, 175, 206
  in half-plane, 132, 134
Borel measure, 1
  set, 1
Boundary, for function algebra, 172
  Šilov, 173, 179
Boundary-values, non-tangential, 34
  of harmonic function, 38, 79
  of H^p function in half-plane, 128
  vanishing of, 52, 58
Bounded characteristic, function of, 73
Bounded variation, function of, 5, 26, 71

$C(S)$, 6
Cauchy integral formula, 28
  kernel, 28
Cesaro means, 16, 23
  summability, 17, 19, 20
Closed set of vectors, 12
Commutative Banach algebra, 90, 205

Complete set of vectors, 12
Complex homomorphism, 92
  representing measure for, 181, 207
  support set for, 181, 207
Conjugate harmonic function, 27
  boundary-values of, 79
Conjugate Poisson kernel, 31, 78
Conjugate space, 7
  of $C(S)$, 7
  of H^p, 137
  of L^p, 7
Convolution, 21, 25
Corona (ϕ), 164

Derivative, 4
Dirac delta measure, 21
Dirichlet algebra, 54, 101, 156, 181, 184
  kernel, 25
  problem, 32

Essential sup norm, 6
Extreme points, 136, 145, 156
  of unit ball in H^∞, 138
  of unit ball in H^1, 139

Factorization for H^p functions, 67
  in half-plane, 132
Fatou's theorem, 34, 123
Fejer's kernel, 17
  theorem, 20
Fiber, 161, 171
Finite measure, 2
Fourier coefficients, 13, 15
  of measure, 20
Fourier series, 13, 15
  Abel summability of, 32, 33, 34
  absolute convergence of, 71, 96, 97
  Cesaro means of, 16
  Cesaro summability of, 17, 19, 20
  of measure, 20
  partial sums of, 15

Fourier transform, 113
Fubini's theorem, 3
Function, absolutely continuous, 71
  analytic, 27
  Baire, 2
  harmonic, 27, 38
  inner, 62
  integrable, 2
  of bounded characteristic, 73
  of bounded variation, 5, 26, 71
  of class H^p, 39, 121
  outer, 62, 103, 120, 133, 139
  singular, 67, 68, 133
  subharmonic, 62
Function algebra, 54, 101, 144, 181, 184, 207

Gelfand-Mazur theorem, 91
Gelfand representation, 159
  for H^∞, 159
  for L^∞, 169
Greatest common divisor, 85, 101

H^p, 39
  conjugate space of, 137
  extreme points of unit ball, 138, 139
  isometries of, 147, 148
$H^p(d\mu)$, 55, 58
$H^2(K)$, 108
Hahn-Banach theorem, 8
Half-plane, H^p of, 121
  of lattice points, 54
Hardy class, 39, 121
Harmonic conjugate, 27, 79
  function, 27 (boundary-values, 38)
Herglotz's theorem, 34, 40, 67
Hilbert space, 9
Hölder's inequality, 3
Hull, 89, 205

Ideal, closed, 83, 85
  maximal, 82, 87, 92
  primary, 83
  principal, 88
  theory, 82
Inequality, Bessel, 11
  Cauchy-Schwarz, 9
  Hölder, 3
  Jensen, 51, 52, 56, 61, 185
Inner function, 62, 99, 175
  part, 68, 69
  product, 8

Integrability of log $|f|$, **52**
Integrable function, 2
Interpolating sequence, 194
Interpolation theorem for H^∞, 203
Invariant subspaces:
  for Dirichlet algebra, 102
  for H^2 of half-plane, 107, 113
  for right translations, 114
  for shift operator, 99, 111, 115
Isometry, 108
  of function algebra, 144
  of H^∞, 147
  of H^1, 148
Jensen's inequality, 51, 52, 56, 61, 185
Jordan decomposition theorem, 5

Kerenel, Cauchy, 28
  conjugate Poisson, 31
  Fejer, 17
  Poisson, 30, 123
  positive, 17
Krein-Milman theorem, 137

$L^\infty(d\mu)$, 6
$L^p(d\mu)$, 3
Lebesgue, decomposition theorem, 4
  dominated convergence theorem, 3
  measure, 2

$\mathfrak{M}(B)$, 159
Maximal ideal, 82, 87, 92, 158
Maximal ideal space, 159, 205
Maximality, 93, 190, 194
Measurable set, 3
Measure, absolutely continuous, 21, 44, 46, 47, 55
  analytic, 39, 47, 51
  Baire, 1
  Borel, 1
  complex, 5
  derivative of, 4
  Dirac, 21
  finite, 2
  Fourier series of, 20
  Lebesgue, 2
  multiplicative, 55
  periodic, 24
  Poisson integral of, 33
  positive, 1
  real, 5
  regular, 1

Measure (*cont.*)
  singular, 46, 55, 66
  zero, 3
Mutual singularity, 4

Non-negative harmonic function, 38, 134
Non-tangential limits, 34, 41, 78, 123
  vanishing of, 52, 58
Norm, 5
  essential sup, 6
  L^p, 6
  sup, 6
Normal family of functions, 189
Normed linear space, 5

Orthogonal complement, 10
  projection, 11
  set, 11
  vectors, 10
Orthonormal set, 11
Outer function, 62, 103, 120, 139
  in half-plane, 133
Outer part, 68, 69

Paley-Wiener theorem, 103, 104, 113, 131
Parallelogram law, 9
Partial sums, 15
Periodic measure, 20
Plancherel theorem, 131
Poisson integral formula, 30
  kernel, 30 (for half-plane, 123)
Positive measure, 2
Privaloff's theorem, 58
Projection, from L^p to H^p, 149, 154, 155
  orthogonal, 11
  theorem of M. Riesz, 151

Radon-Nikodym theorem, 4
Regular Borel measure, 1
  family of functions, 189
Representations of H^∞, 116

Representing measure, 181, 207
Riemann-Lebesgue lemma, 22
Riesz, F. and M., theorem on measures, 47, 51, 56, 59, 72, 186
Riesz, M., theorem on projections from L^p to H^p, 151
Riesz-Fischer theorem, 14
Riesz representation theorem, 7
Runge's theorem, 95

Shift operator, 98, 111
  invariant subspaces for, 99, 111, 115
Sigma ring, 1
Šilov boundary, 173
  for H^∞, 174, 179
Simple function, 2
Singular function, 67, 68, 133
Space of complex homomorphisms, 159
  of maximal ideals, 159
Spectrum, 91
Subharmonic function, 62
Summability, Abel, 32, 33, 34
  Cesaro, 17, 19, 20
Sup norm, 6
Support set for homomorphism, 181, 207
Szegö's theorem, 49, 56, 184

Topology, weak, 159
  weak-star, 8
Total variation, 5, 7

Unitary operator, 108, 134

Variation of measure, 5, 7

Weak topology, 159
Weak-star topology, 8, 156
Weierstrass approximation theorem, 26
Wiener's inversion theorem, 96

X^*, 6

ST. MARY'S COLLEGE OF MARYLAND
ST. MARY'S CITY, MARYLAND
41224